WIFI HOME NETWORKING

WiFi Home Networking

Raymond J. Smith

McGraw-Hill
New York Chicago San Francisco Lisbon
London Madrid Mexico City Milan New Delhi
San Juan Seoul Singapore Sydney Toronto

The McGraw-Hill Companies

Cataloging-in-Publication Data is on file with the Library of Congress.

2 3 4 5 6 7 8 9 0 DOC/DOC 0 9 8 7 6 5 4 3

P/N 141254-9

PART OF ISBN 0-07-141253-0

The sponsoring editor for this book was Judy Bass and the production supervisor was Sherri Souffrance. It was set in Century Schoolbook by MacAllister Publishing Services, LLC.

Printed and bound by RR Donnelley.

This book is printed on recycled, acid-free paper containing a minimum of 50 percent recycled de-inked fiber.

The love of Raymond's life, his Jade—
The brightest jewel that heaven ever made.

CONTENTS

ACKNOWLEDGMENTS

The author would like to thank the following people for their contributions toward making this book possible:

Diana Ying, Public Relations for Linksys (always first with the most). Also, the Linksys' Tech Support team, including Raul Partida, Joe Catindig, Carlos Angulo, Carlos Coral, and Marcelo Deavila.

James Little, PR Manager for Belkin Components.

Winona Man, Sales Executive for Gigafast Ethernet.

C. Brian Grimm of Wavecoms, the Wi-Fi Alliance publicity firm.

Nathalie Welch, Public Relations for Apple Computers.

Megan Behrbaum of Starbucks Coffee.

All who volunteered to be interviewed on the Future of Wi-Fi in Chapter 8:

Christian Gunning, Product Management Director for Boingo Wireless.

Anthony Townsend, NYCwireless and the Taub Urban Research Institute.

Dennis Eaton, Chairman of the Wi-Fi Alliance, who also volunteered for the Foreword.

Gemma Paulo, Wireless LAN market researcher, In-Stat/MDR.

Jim Harrington, Wireless Division Manager for Linksys.

Dave Russell, Apple Computer's Director of Worldwide Product Marketing for Mobile Consumer and Wireless Products.

David Chapin, Hewlett-Packard's Home and Small Office Product Marketing Manager.

Photographic help:

Heather Wilson and Rob and Sarah Edison for providing a photo model.

Anthony Loder, son of the actress/inventor Hedy Lamarr, for a photo of her.

Donovan Brashear and Bill Sutherland of Advanced Network Services for their photo help.

Bryan Jay Miller, the Internet savant for WOXY Radio in Oxford, Ohio.

The Umholtz family, including Ben Umholtz, Faith Curtis, Carson Land, and Jacob Curtis.

Hunter Freeman Photography, courtesy of Apple Computers.

Scott Jordan, president and CEO of SCOTT eVEST, LLC.

Rick Ehrlinspiel, president of SurfandSip.

Tony Stramandinoli, director of marketing, SMC Networks, Inc.

Chapter 2 statistics:

John Horrigan and Lee Rainie. The Broadband Difference: How Online Americans' Behavior Changes with High-Speed Internet Connections at Home. Pew Internet and American Life Project, June 2002.

Amanda Lenhart, Mary Simon, and Mike Graziano. The Internet and Education: Findings of the Pew Internet and American Life Project. Pew Internet and American Life Project, September 2001.

Nikita Borisov, Ian Goldberg, and David Wagner of the Computer Sciences Department of the University of California at Berkeley.

Literary assistance:

Ed Pharr of DeVry Technical Institute for the will, and Tom Rugg of GTE California for the way.

David Leathers of Eye Square Productions, Los Angeles, for the original idea and his encouragement and advice.

William L. Utter of WMUB, who taught me how much can be learned when you shut up and listen.

For the supplemental CD-ROM:

Jon Allen and Adam C. Latham, Intersil.

Kevin Beagley, Hewlett-Packard's public relations manager for Wireless and Emerging Technologies.

Michelle Nash, of CyberPower Systems, and technicians Pietro Boggio and Ryan Jessen.

Bonnie Cheong, vice president of Corp Development and Communications, Gemtek Technology.

Lauren B. Tascan, vice president, S&S Public Relations, Inc. (Actiontec).

Jon Asahina, vice president of marketing, Troy Wireless.

The McGraw-Hill Team:

Judy Bass, editor.

Karen Schopp, West Coast senior account executive.

FOREWORD

When I first heard of wireless computer networking in 1998, I scratched my head and wondered, "Why would anyone want to do *that*?" At the time, radio-attached terminals were restricted to special niches such as warehousing, inventory control, or medical applications that required mobility. One vendor's equipment wouldn't work with another's, and by today's standards, it was terribly slow. I had no idea in those days that the industry would expand into what it has become.

Since then, wireless Ethernet networking has evolved from a niche into a necessity. Affordable Internet access has become widely available to homes and businesses thanks to DSL and cable modems. Laptops have increasingly become a standard tool for doing business. Today's users are more accustomed to having a computer at hand wherever they may be, perhaps literally in hand, with the recent development of PDAs and the steady proliferation of cell phones. Wi-Fi is a logical extension of that. It has brought users ever closer to the data they crave, as our economy and society become more and more information oriented. Wi-Fi has changed the way we do business. It will change the way we live.

In 1998, I could never have guessed at my role in this process. I feel lucky to be part of a large industry of talented people building a technology that is improving the lives of so many around the world. The Wi-Fi Alliance is privileged to make an important contribution to this process. We help our member manufacturers to cooperate as well as to compete effectively with each other. As a result of this "co-optition," wireless computing products are continually becoming less expensive, more secure, and easier to install and use. They are rapidly approaching the ideal point in the user experience, the point at which networking becomes powerful and universal, but also unnoticeable and taken for granted.

The future is even more exciting. Wi-Fi is finding its way into familiar consumer products such as PDAs and cell phones. The same process will occur in your home as wireless speed increases and its potential becomes realized in uses and products we can't imagine today. Wi-Fi is an ideal medium for streaming audio and video out to mobile, handheld view screens, or from your set-top box to the various TVs in your house, for example. All *without wires*.

Your ability to make this technology work for you will be very much enhanced if you first know how it works. When residential high-speed Internet service via cable or DSL is combined with a home network, all your family members can share it at the same time. This book explains both steps: getting the access and distributing it without drilling holes in the walls.

In addition to explaining the technology in terms that the unfamiliar can understand, this easy-to-read tutorial also contains many useful tips that can make installing and using a wireless network in your home an enjoyable and rewarding experience. That's good for you and good for our member manufacturers, who want repeat customers.

In just the past few years, wireless Ethernet has emerged as the medium of choice for users who roam. With a Wi-Fi-equipped laptop (now a standard feature built in to many laptops), you can stay connected to the Internet as you wait for your flight at the airport, as you sip your coffee at your favorite coffee shop, or in your hotel room. Wi-Fi is keeping the promise of nomadic, IP-based, high-speed communications in residential and public places. Wi-Fi is becoming part of the air we breathe.

Smith's book does a good job of explaining the benefits and the steps necessary to extend the World Wide Web to the place where you spend the most time: your own house. There is no longer any reason that your computing experience has to be any different there. We've chosen to participate in *WiFi Home Networking* to further that goal.

We often end our briefings with the motto: Wi-Fi Everywhere! The primary reason you should have a Wi-Fi network in your own home is to make it like everywhere else.

—Dennis Eaton

Dennis Eaton is the current chairman of the Wi-Fi Alliance, an industry association whose 185 members cooperate to certify the interoperability of wireless equipment based on Wi-Fi (IEEE 802.11) and to promote Wi-Fi as the global *wireless local area network* (WLAN) standard across all market segments. In addition, he is a senior strategic marketing manager for Intersil, an Alliance member corporation based in Milpitas, California. Dennis has over 16 years' experience in RF design. He holds a BSEE degree from Michigan State University and an MBA from the University of Central Florida. Mr. Eaton is also an active participant in the Institute of Electrical and Electronics Engineers, a worldwide professional organization responsible for technical standards governing telecommunications and computer equipment.

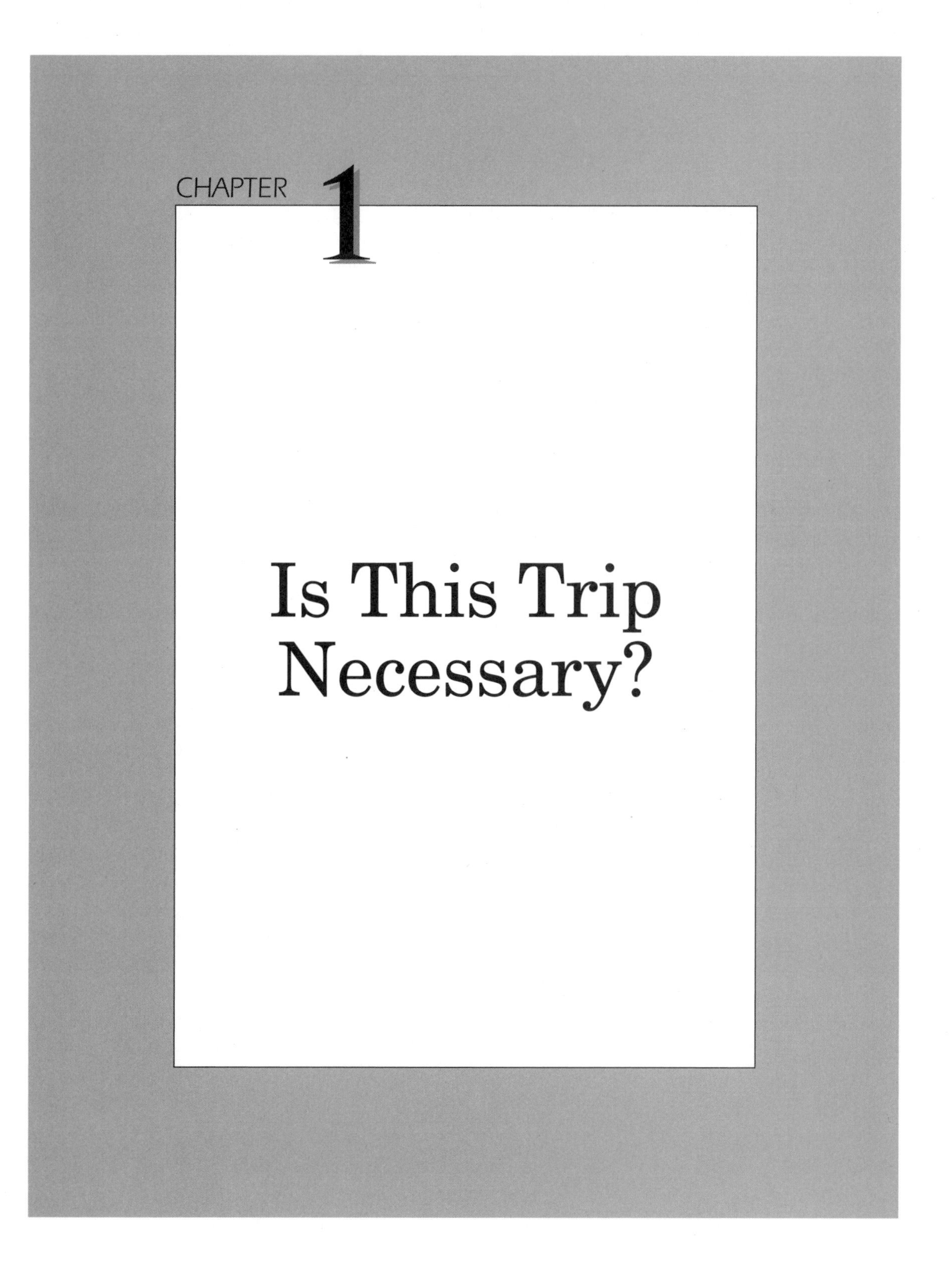

CHAPTER 1

Is This Trip Necessary?

Chapter 1 is a general overview of the costs, benefits, and necessity of data networking. It summarizes the latest, user-friendly hookup methods, particularly ones that do not require drilling holes.

At Home in the Twenty-first Century

Beaver Cleaver wouldn't recognize most of the places we live in today. The main character of that black and white 1957 TV show "Leave It to Beaver" lived in a mythical suburb of Cleveland, Ohio. Beaver's dad was the quiet-spoken Ward. Ward was a rock-stable professional who put on his suit and tie and commuted to his nine-to-five job downtown each weekday as precisely as a mantle clock. (He wore the tie at the dinner table, too.)

Every morning, Beaver's mom, June, would kiss Ward goodbye and stay behind to do the housework. When she was not serving milk and cookies, she would push a vacuum cleaner around the living room while wearing a $50 dress and pearls.

If all that seems like life on another planet to you, it's because the home and its functions in our lives have evolved considerably since then. The suburban split-level with the picket fence and lawn is still there for those who want it. But a vast array of choices is at hand for those of us who want or need something else. Similarly, our economic and social lives have new and practical alternatives.

It's not hard to imagine how the Cleavers would live today. First, old Ward wouldn't have to drive to work every day, nor arrive at a fixed time, nor stay for just eight hours. He wouldn't work at the same company for 30 years. He might work for several at the same time or work for himself as a consultant. Modern Ward would work wherever and whenever he wasn't doing anything else, which means working in airport lobbies, on the commuter train, or perhaps on the kitchen table. You can forget the necktie; sometimes Ward would work in his pajamas. You might see Ward walking on the golf course, talking aloud to no one as a crazed homeless person might, but you should look again. He'd probably be using his cell phone earplug with its built-in wire mike. Ward could be cutting a deal while slicing the ball.

The June Cleaver of 2003 would work, at least part time. June might carry a briefcase or perhaps even wear a tool belt, but whatever she does, her personal computer wouldn't be more than a minute's walk away. She may shop on eBay and probably hasn't seen her banker face to face in months. She might not know the people across the street, but she would be an active member of a community that stretches the breadth of her life experience, unbounded by geographical borders. June could keep track of hubby, the kids, and Gramma from her desk by free, fast email. Her *personal digital assistant* (PDA) would herd her stocks for her and squawk when something unpleasant happened on Wall Street. As with Ward, June's work, her leisure, her family, and her social life would gradually become homogenized into an undifferentiated time stream.

Twenty-first-century Beaver wouldn't haul many textbooks to and from Cleveland Elementary. His most important book would be his Power Book. He could use it to get his assignments, write his papers, pull his research, paint his pictures, take his tests, and fetch his grades. He probably grew up with friendly, powerful, cheap computers. He would be as comfortable with his computer as the original Beaver was with his bicycle. Modern-day Beav would see his laptop as his telephone, record player, and television. It would be a major part of his persona and his gateway to the world. If the Beaver should get grounded, for example, he might use instant chat to stay tight with Eddie Haskell and other friends. He could play Saucer Command in real time with e-buddies in another state. He might even feel closer to some of his e-pals than to his own big brother, Wally.

Thanks to the World Wide Web, we are all elbow to elbow in cyberspace. As a people, we are becoming more and more reliant on instant information and communication to carry on our lives. Those without access to data cannot compute, at least, not for long. And those who cannot compute cannot compete. Thanks to the sudden arrival of wireless home networking, also known as *wireless fidelity* (Wi-Fi) or *wireless local area network* (WLAN), fewer and fewer "dead spots" exist where that ocean of data is not available at a nearby faucet. Often, one of those places is the contemporary home. This book will tell you how to make your house or apartment into a connected, on-the-air hot spot.

Turning Your Home into a Hot Spot

The benefits of networking computers are so obvious that businesses today accept the necessity as a given expense and standard practice. Perhaps the best example of those benefits, and the problems you might face today as you network your home, is the first wireless network. The wireless network was born in 1970 and called *AlohaNet,* because it linked the Hawaiian

Islands. Norm Abramson, a professor who relocated from Stanford's engineering department to the University of Hawaii, designed AlohaNet. He explained that the surfing—real surfing, not net surfing—was better there.

A Brief History of Wireless Computer Networks

Hawaii presented a unique challenge and opportunity for those engineers and graduate students. At the time, computers were capable of being linked, but only across a computer room floor using a cable the diameter of your thumb. The only alternative was known as Sneakernet, in which data was put onto a floppy disk or tape and carried to a distant computer, presumably by someone wearing comfortable sneakers. That was inadequate for the islands, which were separated by distance and natural barriers. Moving a spreadsheet by courier might entail an expensive boat or helicopter ride. The tapes or printouts could get lost or damaged, and in any case the information they contained would be aging while it was in transit. Multiple copies would accumulate, and corrections or updates required more boat rides.

Those limitations are actually familiar to today's commuters, who often carry their CD-ROMs or laptops from their offices to their homes and back again day after day. It would be so much easier if they could just move the information and leave the battery packs, wires, and floppies behind.

But Hawaii's isolation in the middle of the Pacific also handed the AlohaNet engineers a solution. Plenty of radio frequencies were available for use as transmission media. The islands consisted of relatively few people and therefore relatively little man-made interference. The flat ocean provided unobscured line-of-sight paths for the signals to travel. Soon a blizzard of data was flying from Lanai to Oahu at the speed of light. The concept had been proven, but could it be exported to the mainland? The answer was: not easily.

Coaxial Cables California abounded in mountains, which do not pass radio waves. Wherever you tuned on the radio dial in metropolitan areas, someone was already there. When a new frequency band became available through regulation or a technology advance, lots of entrepreneurs waited to fill the vacuum. Wireless networking had to be put on hold as engineers were forced to return to *coaxial cable*. This wire used a central element surrounded by a layer of insulation and a cylindrical metal or woven shield.

It was expensive because it had to be manufactured to precise tolerances. But data signals did not escape from it to pollute the airwaves, nor did interference penetrate. The scheme was given the name *broadband*, because as the amount of digital data increased, it could be apportioned onto additional frequency channels. Often those channels were shared with traditional video signals on the same type of cable that brings television signals into your home.

Multiconductor Cables When cheaper technologies were developed a decade ago, the coaxial cable was deemed obsolete and replaced with the *local area network* (LAN) drops we know today. These are unshielded *multiconductor* cables, about the diameter of a chopstick, adapted from use by the telephone industry. Their immunity from signal pollution comes from the arrangement of the wires inside. To this day, *universal twisted pair* wiring remains the most common way of making computers talk to one another.

It is rare when a technology rises from the dead, but propelled by an unquenchable popular demand, broadband TV cable is once again being used to deliver data to millions of homes. Likewise, recent advances and refinements have revived wireless as a means of easily interconnecting computers inside homes, businesses, and home-based businesses. That realm has been given the acronym SOHO, for *Small or Home Office*. Over a hundred manufacturers are sending SOHO-targeted products to market.

The Internet Arrives As the means of interconnection has evolved, so has "the end." The Internet began with the National Science Foundation Act of 1950, as an A-bomb-proof means for academics and military researchers to send text to one another. Until 1992, using it for profit was actually illegal. Since then it has become a commercial, all-encompassing medium unto itself, augmenting and even displacing traditional communication channels, such as radio, television, telephones, newspapers, and even the postal service. Some new businesses, such as Amazon.com or the Google search engine, could not exist without it.

At one time, home telephones were considered to be either a luxury or a business imposition. But they quickly turned into a necessity, such that most homes now have several, with many on second or even third home telephone lines. In turn, these wired phones are being augmented or displaced by cell phones. The same process is happening with home computers, as they migrate from the business desktop to the kitchen tabletop. Homeowners know that their PC power is multiplied when they share information.

DSL Arrives The stage was set for Wi-Fi home networks with the arrival of *digital subscriber line* (DSL). This development enables your connection to the telephone or cable TV systems to double as high-speed data channels to the Internet. To get the 1.5 Mbps speeds that DSL now offers, customers previously would have to purchase a dedicated T1 line that would cost between $700 and $1,100 per month. A DSL connection, which can carry regular phone calls as well as data, costs about a tenth of that. Like broadband cable, telephone infrastructure has the considerable advantage of already being in place under streets and hanging from telephone poles. You will not need to dig a trench across your front lawn to get connected.

All these technologies can now be applied in combination so that the Internet or the World Wide Web can finally cross the last mile into apartments, shops, and split-levels from Cleveland to Fairbanks. Practically anyone can afford it. As time marches on, few can afford to ignore it.

Wirelessly Wired

According to *International Data Corporation* (IDC), about half of all American households now have a computer. More than 20 million of them have more than one. Most purchases of personal computers are not by first-time buyers, but by those who already have one. IDC also estimates that by 2004, some 4.2 million U.S. homes will have Wi-Fi networks. But you may be asking yourself, "Since I've gotten along fine without a network at home, why do I need one now?" A Wi-Fi home network will enable you to

- Share a single high-speed Internet connection and *Internet service provider* (ISP) among several PCs at once.
- Feed one printer from several PCs.
- Exchange pictures, audio files, and spreadsheets among all your PCs.
- Back up data between PCs.
- Remotely control one PC from another.
- Play games against other players on other PCs in the house or in other locations.
- Work anywhere around the house.
- Send live video from your backyard.
- Use your laptop without having to plug anything in.
- Work at home as you do at work or make your home-based business competitive.

Because you have purchased this book, you probably already have some problems you want solved. Perhaps you are tired of arguing with your spouse and kids who are waiting in line to log on. Or perhaps you want to be in on the next wave or be the first on your block to own your very own gee-whiz mobile network. Perhaps your boss has suggested you take work home like the other employees are doing, but after using the company's high-speed LAN, the home phone modem is positively painful. And maybe you've had a second phone line installed for it, so that you won't be cut off from voice service for hours at a time.

The Old Way of Connecting Dial-Up Modems

Until recently, home-base workers and surfers were stuck with clunky dial-up telephone *modems*, which stands for modulator-demodulators. Modems change a computer's internal data into sound and back again, so that it can be sent over ordinary telephone lines. The process of using an audio channel to carry digital information is inherently inefficient. A modem is pictured in Figure 1-1.

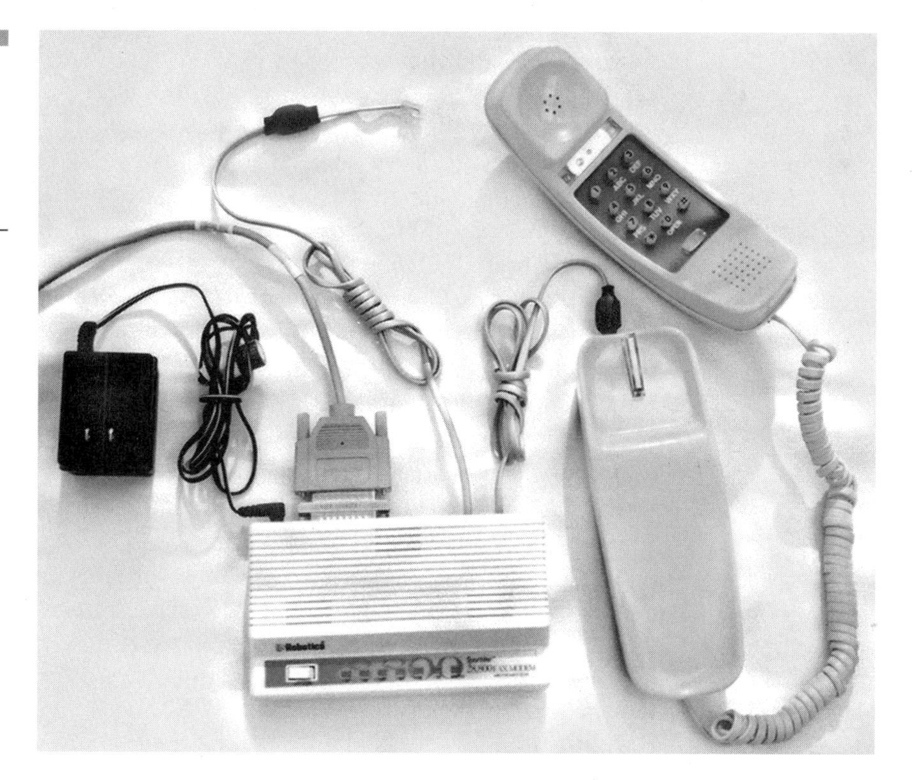

Figure 1-1
A standard modem, the old way of connecting to the Internet. Photo by Daly Road Graphics.

Top Ten Reasons to Hate a Phone Modem

1. Modems are complicated to set up. Wires sprout from the wall to the modem and from the modem to the PC. You need multiple dial-up numbers, scripts, serial port parameters, and so on.
2. Modems are not on all the time, like the connection at work.
3. It takes too long to make it connect to the Internet.
4. A modem makes funny noises every time it connects. Sometimes it doesn't connect and you have to start over.
5. While it is connected, nobody else can use the phone or fax.
6. While someone is talking on the phone, you can't connect.
7. When the modem is connected, only one PC can use it.
8. It is slow. If the telephone line is noisy, it gets very slow.
9. Modems tend to hang up from time to time unpredictably.
10. Did we mention the part about being slow?

Wiring at Home and Work

The standard workplace network connection uses a format called *Ethernet* carried on universal twisted pair wiring. It moves data at a rate of 10 Mbps under ideal conditions. Nationwide, this is quickly being upgraded to the aptly named *fast Ethernet*, which goes 10 times as fast at 100 megabits. In practical terms, this means that downloading a new copy of Internet Explorer at work will take five minutes or less. The fastest modem connection available is 56 *thousand* bits per second tops, and usually less. At home, that same download would take five *hours*, assuming that your phone connection will stay up continuously for that long. A modem connection might be okay for picking up email, but web sites that flash onto your screen at work will squeeze out like toothpaste over a phone line.

As time progresses, more web pages are being designed with the assumption that the viewer has a fast connection. Webmasters are caring less and less about *bandwidth*, or user capacity, and more about fully utilizing the *graphical user interface* (GUI), to impress their viewers. Animated graphics and musical accompaniment are turning web sites into multimedia experiences at the cost of transmitting long files. Some applications, such as live teleconferencing, emailing pictures of kids to relatives, or sending music files are all nearly impossible through such a skinny data straw (see Figure 1-2).

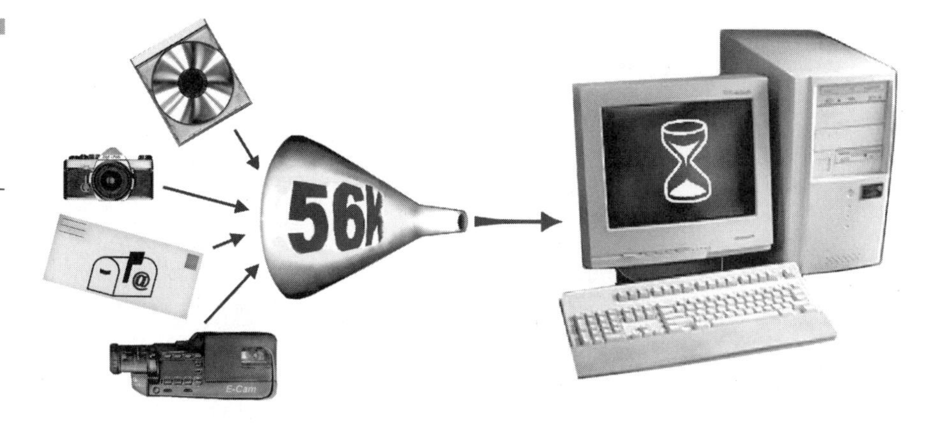

Figure 1-2 The clog of information squeezing through a modem connection

Aside from the historical lack of a high-speed path to the Internet, a primary reason that families have not installed networks in their households is the "hassle factor" associated with cable. LAN cable does have advantages. As mentioned, it runs faster and it is also more difficult for a snoop to tap into. But those of us who've worked in offices have seen the great effort required to string lines from a nearby electrical closet to every desk in a room. It can take days for a two-person team, even with a raised computer floor in a building designed to allow for it. If the office grows or relocates, then the work must start again. And if the office downsizes, the IT manager cannot recover the budget that has been paid for excess capacity.

Many new homes have data cabling built in, just as telephone, electrical, and, more recently, cable TV lines come preinstalled. That's evidence of how quickly home networking has become taken for granted. (In Chapter 3, "From the Doorstep to the Tabletop," we will describe home network schemes that use those same telephone and power lines.)

Computer cable, also known as *category 5* (CAT5), is cheap and relatively easy to run through support studs before the covering drywall is up and painted, especially when it is bundled with the other wiring. But *retrofitting* is much more problematic. Even with long, spring-steel drill bits specifically designed for the task, you face the danger of puncturing a pipe or cutting a power line. (The author has done both of those things.) Cinderblock or poured concrete walls are particularly difficult to deal with. The lines must be kept away from sources of strong electromagnetic interference, such as fluorescent lights and power outlets. Coiling up loose cabling inside a dropped ceiling, for instance, can also cause the cable to pick up interference.

When it is possible to run cable in the crawlspace under a home, the homeowner is forced to, well, *crawl* amid the bugs, dust, and claustrophobic darkness. For apartment dwellers, punching holes in walls they do not own may be grounds for their eviction.

In desperation, some users have been known to staple cables around baseboards and over doorways. Cables can also be found sandwiched under their rug, or even on top, "protected" with an ugly strip of duct tape. These methods may suffice for a garage or a dormitory room, but would never win the Martha Stewart seal of approval in a living room or den. Once the LAN sockets have been mounted on the walls, there remains the problem of how to connect them all together. And then how do you connect all those connections to the Internet?

Wi-Fi requires some work also, of course. But the process is much more manageable, and the payoff is clearly greater in terms of ease of use. If you prefer a PC, you will be happy to learn that Internet connection sharing has been bundled into the Windows operating system, and built-in Wi-Fi self-configuration is a powerful feature of the new Windows XP and Windows 2000. Mac users have not been left behind. The Apple AirPort (see Figure 1-3) enables them to get their Macs on the air to share Wi-Fi hardware and swap data with Windows PC users over the same home network.

Interchangeable Parts

Eli Whitney may be remembered as the inventor of the cotton gin, but one of his other inventions led the way to mass manufacturing and the popularity of Wi-Fi. He designed rifles for the Union Army that could be disassembled so that the pieces could be used to build or repair identical rifles.

Figure 1-3 Apples and PCs can talk using Apple's Wi-Fi AirPort. Photo by Hunter Freeman Photography, courtesy Apple Computer.

The key to the utility of Wi-Fi is its adherence to a global standard. Like the Internet or the Java programming language, it works across hardware platforms and network operating systems. You can put your network together using pieces from a dozen different manufacturers interchangeably. You can carry your PC from room to room, house to house, or hotel to coffee shop, and not have to worry about the brand of access point in use at those places. One verification to this approach is the fact that similar industry consortiums have formed to standardize home telephone and power-line networks.

The *Institute of Electrical and Electronics Engineers* (IEEE) hears proposals, requests, and arguments from manufacturers, scientists, and students and then develops a technical compromise, or standard, that all or most can live with. Their wireless Ethernet standards are detailed descriptions of how an over-the-air interface between a client and base station, or between two clients, must operate. It defines signal strength, data format, and speed. The standard is also a recipe for encryption, so that unintended recipients cannot read someone else's data.

IEEE Wireless Standards

The first attempt at a wireless standard was the *HomeRF* protocol, which did not catch on because of its slow (1.6 Mbps) speed. It was replaced by the 802.11 standard, which ran at 1 or 2 Mbps. Because of its limited speed, it is also history. In 1999, the IEEE added the "a" and "b" refinements.

Products conforming to the 802.11a standard operate at speeds of up to 54 Mbps on a very short-wave frequency of 5 billion cycles per second, or 5 gigahertz, abbreviated GHz. Its speed advantage is offset by its shorter range, which is typically 50 to 200 meters. Unlike the more popular b standard, it uses a modulation scheme with the hefty name of *Orthogonal Frequency Division Multiplexing* (OFDM) that makes possible data speeds as high as 54 Mbps and cuts down on cross-channel and reflected-signal interference. More commonly, communication takes place at 6, 12, or 24 Mbps.

Today the most widely followed standard by far is known as IEEE 802.11b. It moves data at a top speed of 11 Mbps in the 2.4 GHz frequency band. It is more prone to interference than 802.11a, but the lower frequency gives it a longer range, estimated at between 75 to 300 meters.

The 802.11g standard is now under development and discussion by members of the IEEE, but it won't become official until released in 2003. The g standard will be backward compatible with the b standard and will operate on the same 2.4 GHz frequencies. But it will be faster at 54 Mpbs and less vulnerable to radio noise. Its greater capacity makes it a promising media for wireless streaming video.

As often happens with a burgeoning technology, as soon as the Committee handed down the standard, it began amending it. Other additions are in the IEEE pipe:

- 802.11d and h will accommodate European regulations governing radio devices.
- 802.11e is due in January 2003. It adds *quality of service* (QoS) features.
- 802.11f will add protocols that enable data sharing between disparate systems in 2003.
- 802.11i addresses security holes in the present standards.
- The *Wireless Next Generation* (WNG) specification seeks to combine all the above into one universal standard.
- *Ultrawideband (*UWB) was granted a limited license in February 2002 for use in the 3.1 GHz and 10.6 GHz bands, but only indoors or in handheld peer-to-peer applications for now.

These are explained in greater detail in the Chapter 2, "From the World to Your Door."

Nonstandards

As if that alphabet soup were not confusing enough, some manufacturers are not inclined to wait for an officially sanctioned speed jump. They want to have it both ways by introducing models that will fall back to a standard rate when talking to equipment built by another manufacturer, but go faster between themselves. D-Link, for example, has a B version that runs at a nonstandard 20 Mbps, but slows to the standard speed of 11 Mpbs in a multivendor environment. Similarly, Linksys advertises a turbo mode of 72 Mbps between its A interfaces, and Cisco has a similar option. In order to remain standard, they must slow to 54 Mbps when talking to other manufacturers' A equipment. The higher turbo speed also implies shorter ranges, to perhaps five meters between stations.

The IEEE does not have the legal authority to enforce any of these standards in the manner that the *Federal Communications Commission* (FCC) does. In fact, Wi-Fi operates on an unlicensed portion of the radio spectrum, designated as the ISM, for *Industrial, Scientific, and Medical* band. This is good because you don't have to apply for a license. It's bad because microwave ovens, cell phones, baby monitors, and other users are in the same pool with you. If these bandwidth competitors trample on your signal, you have no official authority to complain to. As the number of Wi-Fi technology users jumps upward from 20 million in 2002, interference is bound to become a common problem. Crowding will promote the expansion into other frequency bands. The 802.11a equipment runs in the quieter 5 to 6 GHz *Unlicensed National Information Infrastructure* (U-NII) spectrum.

Manufacturers are already designing equipment that will automatically shift from a to b to g, or even interface them all inside one box. This way, when the new equipment arrives, it will be backward compatible with the hardware on the shelves today.

Wi-Fi Versus Bluetooth

Another wireless technology is on the market today, which is similar enough to Wi-Fi at first glance to add to the confusion. The Danish telephone equipment company Ericsson originally developed *Bluetooth* (BT) *wireless*. (Ericsson named their radio technology after a tenth-century Viking king who had a sweet tooth for blueberries.) BT uses the same radio band as Wi-Fi and a similar frequency-hopping scheme to dodge interference. Some application overlap also exists, since some devices using the Bluetooth standard can interface a PC to a printer or to a wired network, just like Wi-Fi. Some equipment is being designed today, which combines interfaces, one for BT and another for Wi-Fi. But other than that special circumstance, Bluetooth and Wi-Fi are not compatible.

When used in close proximity, Bluetooth devices are likely to interfere with Wi-Fi devices. They both employ a frequency-hopping scheme that minimizes fixed-frequency interference, but they both hop among the same range of frequencies, and sometimes hop on top of one another. Moving the BT device can often solve the conflict, since its operating power and range are less than Wi-Fi. Newer BT models, using technical tweaks approved by the FCC, will most likely peacefully coexist with Wi-Fi.

An argument exists in the wireless industry about which of these two technologies will dominate in years to come. But even now the two standards are becoming more differentiated and applied to different customer needs. Wi-Fi is a method for connecting PCs and printers to a LAN over a distance of 75 to 300 meters. Bluetooth, with its shorter 10-meter range, deals with a *personal area network* (PAN). Its unique application is the elimination of wires to handheld or wearable devices, such as PDAs. A wireless headset for a cell phone would be another good example. Unlike a structured LAN, these personal devices link up on an ad hoc basis.

Your Best Bet If all these standards are puzzling, your safest bet remains the 802.11b standard. So many people are using it that it won't be abandoned anytime soon. A manufacturer who wants to prove that his product will work with other manufacturer's Wi-Fi equipment must submit it to the Wi-Fi Alliance for testing. Before you buy a piece of Wi-Fi hardware, you should first examine the package for a prominent label that indicates conformance. Very likely you will find another label indicating which radio band the equipment uses. Another way to verify standard compatibility and interchangeability is to surf the Wi-Fi Alliance web page at www.wi-fi.org/certified_products.asp. Check before you sign the check.

Why Wi-Fi?

WLAN products have established themselves in the marketplace in a very short period of time. The In-Stat/MDR research agency predicts that 8.2 million Wi-Fi units will have been sold by 2003. The surprisingly intense interest shown by consumers and manufacturers alike is an indication that however much the technology may change, it is here to stay (see the Gemma Paulo interview in Chapter 8, "Where No One Has Roamed Before").

In July 2002, industry giant Microsoft announced their intention to jump into the Wi-Fi market. In a press release, Randy Ringer, general manager for the Microsoft Hardware Division, summarized their motivation: "Today, the PC is at the center of productivity in most homes. People use it to communicate with friends and family, browse the Internet, and track their finances. But consumers want their PCs and their electronics to do more than just the 'traditional' tasks; they want to play head-to-head games, talk to and see friends across the country over the Internet, and access and play their music from any PC in the house. That's where the Microsoft wireless networking products come in. They will allow users to access the Internet and their data from all areas of their house, regardless of which computer it's housed on, while adding expanded mobility and the option of putting a PC wherever they want it."

If you take a walk down the aisle of any major electronics store, you are likely to see several brands on the shelves, ready to go. Here are some reasons you should consider Wi-Fi as the means to install or expand your next computer network:

- It's easy to add on to an existing Ethernet network.
- It's easy to expand the number of users and operating ranges.
- It's relatively efficient. Newer versions run even faster.
- Given the low signal power, it is surprisingly reliable.
- It has a good operating range, between 75 to 300 yards, depending on conditions.
- Mix or match. Wi-Fi hardware is interchangeable with other Wi-Fi hardware.
- Different hardware platforms can use it as a common medium of exchange.
- It's relatively cheap, and getting cheaper all the time.
- You can get up, move around, and stay connected.
- Roaming opportunities will make your investment more, not less, valuable over time.
- It's easy to throw together a network on the fly.

Safety The output power of Wi-Fi equipment is low, much less than a cellular phone. For that matter, only one study by a Swedish oncologist, Dr. Lennart Hardell, has claimed a connection between older, analog phones and health. Three major studies published since December 2000, including one by the U.S. National Cancer Institute, found none.

Specifically, the 802.11b Wi-Fi standard enables a maximum output of 1,000 milliwatts, or 1 watt of transmit power. For comparison, a CB radio runs at 4 1/2 watts. Even so, most 802.11b products are limited to 30 milliwatts to avoid overheating and save batteries. The newer 802.11a standard enables between 40 and 800 milliwatts, maximum, depending on what part of the 5 GHz spectrum is being used. Radio waves fade out over distance, so the exposure to users of a Wi-Fi LAN is negligible. Even in a room with 100 wireless nodes, only one will be transmitting at any given moment, and unlike the steady, connection-oriented broadcasts from phones, Wi-Fi equipment sends data in short bursts. The short answer to the safety question is that no nasty health effects have ever been attributed to Wi-Fi.

Roam, Roam on the Range

This book focuses on the use of Wi-Fi in the home where it works so well, chiefly because of the mobility it imparts to the user. As this is being written, that mobility is being multiplied a thousandfold, and with it, the value of your domestic investment. Outfitting your PC for use in a Wi-Fi home LAN will enable it to work in numerous places outside and not only in Wi-Fi business offices. This new empowerment is one of the most exciting aspects of Wi-Fi.

A dynamic synergy is developing between governments, manufacturers, and traditional and new service providers. They are combining efforts and their advantages to propel the Wi-Fi standard into a global phenomenon that is greater than the sum of its parts. Here are some of the latest developments.

Wi-Fi Drivers: The Private Sector

With Wi-Fi, any place where people congregate to wait can be an opportunity to connect to the Internet. Hotels and motels have been providing phone modem and LAN connections to their traveling guests for years, sometimes at exorbitant rates. Many commercial enterprises are setting up Wi-Fi access points as a value-added service to their customers in the same way that some restaurants offer free newspapers. In this way, every café can become a cybercafé without having to invest in hardwired video terminals, LAN drops, AC outlets, and the rest. Customers carry their own terminals into these hot spots or, in the case of the Surf and Sip chain, client shops and hotels will provide Wi-Fi-equipped rentals (see Figure 1-4). The President of Surf-and-Sip, Rick Ehrlinspiel, says that as of mid-2002 his aggregated network had linked about 140 locations nationwide. By the end of that year Surf-and-Sip expected to include between 500 and 600 sites. Ehrlinspiel added that their research estimates a potential market of 15 to 20 million users who will need their kind of service. The future looks extremely good, Ehrlinspiel added.

Other Wi-Fi venues include airport lobbies, industrial parks, and shopping malls. In the gadget-loving nation of Japan, McDonald's announced plans to deploy Wi-Fi in McDonald's restaurants. As of July 2002, the online directory hotspots.com lists over 1,400 locations in America. More are being added to that list every day. The research agency InStat/MDR predicts that approximately 5,000 hotspots will exist at the end of 2002 worldwide and approximately 41,000 at the end of 2006.

Figure 1-4
Surfing and sipping, the Wi-Fi way. Photo courtesy Surf-and-Sip.

Here are some other hotspot guides on the Web:

- www.wayport.net/locations
- www.personaltelco.net/index.cgi/WirelessCommunities (worldwide links)
- www.wifinder.com
- www.boingo.com

As you might expect, Wi-Fi is ideal for school campuses, where students shuttle from class to class every hour. In mid-2002, IBM announced it would use Wi-Fi to connect over a million students in the New York City school system. Dartmouth's 160 buildings are fitted with over 500 Wi-Fi antennas, spreading coverage over their 200-acre campus. Other well-known collegiate hot spots include the University of Texas at Dallas, the University of Minnesota, Carnegie Mellon (known as the Wireless Andrew project), and the University of California at San Diego.

New Service Providers Some public-spirited grass roots organizations, such as NYCwireless, are setting up free public access points called *freenets*, which cover city parks and streets (see the interview with one of the co-founders of this freenet in Chapter 8.) Like-minded organizations have banded with them to do the same in other cities across the globe.

Larger ISPs equate this access sharing to piracy. But others, such as New York's Bway.Net or AtlasBroadband.com hope to attract more subscribers by offering contracts that allow their customers to give service away to passersby.

Here are some free Wi-Fi resources:

- www.freenetworks.org
- www.pdxwireless.org (Portland, Oregon)
- www.seattlewireless.net (Seattle, Washington)
- www.bawug.org (San Francisco Bay area)

Freenets are fine for net surfers who need nothing more, but business customers require security, reliability, and predictability. To get those features, they must pay someone to provide it, so a new breed of *wireless Internet service provider* (WISP) has surfaced. These specialize in providing standardized Internet access across varied locations. They do so by conglomerating scattered hot spots into a shared billing scheme. One example is Joltage.com, which distributes free software that opens any home access

Wireless Carrier

In March of 2002, Deutsche Lufthansa AG, Europe's second largest air carrier, announced plans to deploy Wi-Fi Internet service on its fleet of 80 aircraft. These airborne networks will be available on routes from Europe to North America and Asia. A Boeing subsidiary, Connexion, will provide satellite data links from the planes to the ground at a throughput rate of 5 Mbps downstream and 750 Kbps upstream. Cisco's 350 Series Wi-Fi equipment (see Figure 1-5), will deliver up to 11 Mbps within the onboard network.

Passengers with laptops or other Wi-Fi enabled devices will be able to access the Internet while in flight faster than with terrestrial DSL or cable modems. The airline will also provide 10 Mpbs wired networks, but they anticipate that wireless LANs will become preferred. Lufthansa plans to migrate to WiFi-only in order to save weight. (LAN cables to every seat on a Boeing 747 add hundreds of pounds and lower fuel efficiency.)

Figure 1-5
Wi-Fi in the friendly skies. Above: Cisco's Aironet Wi-Fi line. Photo courtesy Cisco. Below: A Lufthansa Boeing 747-400. Photo: Astrovision/ Lufthansa.

point to the public. The public access point's owner then gets paid when other roaming Joltage subscribers use it.

Another new WISP is Boingo.com, started by one of the founders of Earthlink. (See the interview with Boingo's director of product management in Chapter 8.) Boingo attracts new customers by offering free software that enables a roaming PC to search for any nearby hot spot, whether or not it is a Boingo subscriber. If it is one of Boingo's 600 outlets, the user is automatically connected and later billed a predictable monthly fee ($75 is typical). A similar California company, iPass, has announced an aggregation deal with five regional European Wi-Fi network providers and one Asian provider. It will allow a subscriber to any one of several overseas networks to roam into facilities owned by the other five.

These nontraditional wireless Internet service providers are actively knitting themselves together so that roaming users won't have to reconfigure their terminals or cough up a credit card number every time they connect. The industry is still waiting for a universal payment method that works across provider boundaries.

Governments Pave the Way The FCC approved new technical refinements this year that make the 802.11b standard faster and more reliable by using multiple signal path technology. New regulatory adjustments will hopefully eliminate most of the interference problems between Wi-Fi and future Bluetooth devices. The FCC so far has declined to restrict Wi-Fi use, as requested by satellite radio broadcasters, who fear interference with their signals.

In early 2002, Phillips Electronics became the first to get approval for its 802.11a Wi-Fi interface in the form of a *Conformité Europeenne* kite mark. This (CE) symbol of approval on a piece of equipment indicates that all pertinent European Community legal requirements have been met. Phillips' card can now be sold throughout the continent because it meets European Union health and safety regulations. This greatly increases the likelihood that America's IEEE A standard, with some add-ons, will also be approved.

In mid-2002, the British Board of Trade rescinded a rule that forbade companies from setting up commercial hot spots. It had an immediate effect similar to the legislation that first allowed profitable use of the Internet. British Telecomm promptly announced that they would initially deploy 4,000 Wi-Fi locations around England.

Hardware Galore Equipment manufacturers are giving their own push. Thanks to Wi-Fi compliance, consumers have considerable choice in whose equipment they buy. That freedom of choice guarantees that competition is intense and prices are kept low. Some manufacturers are cross-marketing with the WISPs mentioned earlier so that you can easily buy certified hardware when you buy access. IBM and Toshiba have both announced their intention to promote hot spot infrastructure, so that customers will have more motivation to buy their Wi-Fi-equipped laptops. WLAN interfaces are included as built-in peripherals on many high-end laptops, Pocket PCs, and Personal Data Assistants today in the same way that optional telephone modems, infra-red ports, and LAN plugs became standard issue in the past. As mentioned, Wi-Fi autoconfiguration is built into the latest Windows network operating systems.

Wi-Fi and Cell Phones

A last barrier to universal Internet connectivity is coming down, and that is the limited range of Wi-Fi transmitters. Cell phone makers have naturally looked forward to the day when their customers could access the Web using their cell data channels. Their solution has been referred to as *third-generation* (3G) technology. Sprint PCS is only one of several companies with 3G rollout plans. The technology also goes by the acronyms *General Packet Radio Service* (GPRS) and *Wideband Code Division Multiple Access* (WCDMA). By whatever name, the service moves data at a comparative snail's pace of 144,000 bits per second. But it does have the important advantage of working wherever a cell phone does, such as in the middle of a cornfield.

Rather than trying to beat Wi-Fi, the cell phone companies are joining in. Ericsson, the previously mentioned inventor of Bluetooth, is one of many who are developing cell phones that will sense the presence of a nearby hot spot. Once acquired, these combination phones will seamlessly divert traffic through that hot spot, using its 11 Mbps data stream for however long it may be in range.

Other nontechnical problems must be dealt with before a universal communicator can be built. Customers will prefer one bill only, regardless of where they've been roaming, and far more cold spots exist than hot ones. But not long ago, when the cellular phone was first introduced, it was a novelty that only worked in a few metropolitan locations. Now ubiquitous cell towers are radiating inside church steeples, atop power transmission towers, and in tall buildings across the horizon (see Figure 1-6). In the same way, it's easy to see a time when Wi-Fi antennas will sprout from street lamps and atop billboards. When this is accomplished, the combination cellular/Wi-Fi terminal will make the world into a continuous hot spot. So another reason for using Wi-Fi in your home network is to make your household like everyplace else.

Wi-Fi Building Blocks

A home network is going to consist of several components in combination: hardware, software, and financial. We will describe them in detail in Chapters 2 and 3, but here is a summary. For convenience, we have broken down the task of getting your home onto the Internet into five steps:

Figure 1-6
A ubiquitous cell phone tower, disguised as a palm tree. The one in the foreground is real. Photo by Daly Road Graphics.

1. Get a link from the Web to your door.
2. Buy Internet access from a provider.
3. Install interfaces on your PCs.
4. Install and configure the software that drives the hardware.
5. Use your home net effectively to get smarter, make money, or have fun.

Connect Your Home

The PC has evolved from a mathematical calculation machine into a powerful communications terminal. Although a home Wi-Fi network can easily connect home-based PCs to each other, an important reason for having such a network is the easy distribution of a wealth of information from the World Wide Web. The Internet is a vast and rich source of entertainment, research, news, weather, sports, and personal contact. Presently, four methods are available for putting it all on tap in your home:

- DSL over cable utilizes otherwise empty radio frequency bandwidth on your cable TV line.
- DSL can also be carried by an inaudible, high frequency audio signal on your existing telephone line.
- Satellite Internet uses the same link as satellite TV, only bidirectionally.
- Fixed-point wireless DSL is essentially long-distance Wi-Fi.

Cable DSL If you have a TV cable in your home, you may have a 2 Mbps data link available to you already. Many cable TV companies, but by no means all of them, set aside a portion of their bandwidth to provide two-way communications like their rivals, the telephone companies. (Ironically, it was the phone companies who first developed DSL in hopes of sending video over their lines so they could compete with cable companies.) As with all the other methods listed earlier, you will have to call or consult the web page of your cable company in order to learn if they have data service in your area. If the answer is yes, you can be online fairly soon.

Cable modems are more standardized than copper wire modems, so you can get a lower install price if you buy one from a store and install it yourself. The process is not that difficult, but they will send help, if you need it, for a price. When you are using your cable modem, you may continue to watch television and use the telephones too.

Telephone DSL Digital service from your telephone company is commonly called ADSL, for *asymmetric digital subscriber line*. Asymmetric refers to the common practice of having two transfer rates: a fast one (typically 1.5 Mbps) for information flow *into* your network (the *downlink*) and a slower *uplink* speed (360 Kbps) for data flow out to the Internet. Most users don't notice the difference, except when they forward a long mail message. (Sending it takes longer than receiving it.) Asymmetric service is cheaper. A variety of speeds are available up to 6 Mbps, and you can order a symmetrical DSL line if you think you need one.

ADSL is the most widely available option today, because not all homes have cable TV systems, but almost all have telephones. Limits exist here also, however. You must be within three miles of a switching office, and the office must be equipped to handle digital service. If you call your telephone provider, they can tell you if you are a candidate and also send a test signal down your line to see if it is ready to carry the digital signal. Not all of them are ready. Some need to be rebuilt, with voice devices such as *bridge taps* and *load coils* removed. Some phone companies are unwilling to spend the

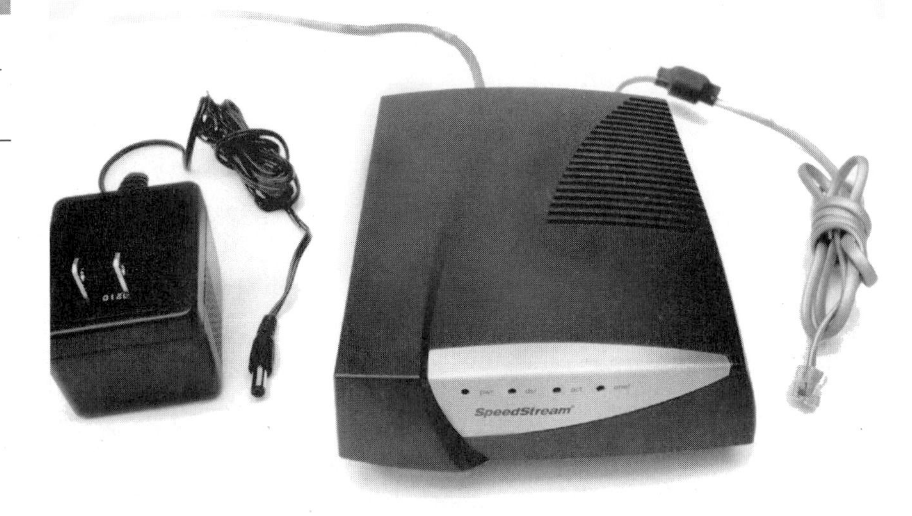

Figure 1-7
A telephone line DSL modem. Photo by Daly Road Graphics.

labor and will simply turn you down as a DSL customer. It may take a month or more for the DSL modem to arrive (see Figure 1-7).

When the DSL modem arrives, you are expected to connect it to your phone system yourself, but that is a simple matter not much more complicated than plugging in another telephone. Then you must add microfilters between your existing phones and the wall outlets to isolate the phones from the digital signal. That is also a simple process, which does not require tools.

As with cable modems, you must have an NIC installed in your PC, and then connect that card to the modem. Software that will introduce your computer to the network must then be loaded onto the PC also, just as in any industrial LAN.

Satellites Like the cable TV companies, two satellite TV companies would like to become major Internet conduits as well. Their satellite channels can be made to carry data in both directions. Transcontinental corporations have been using satellites to shuttle voice and data for years. But their service is comparatively expensive for individual homeowners, because to use it a trained professional must first install a yard-wide parabolic dish (see Figure 1-8). Unlike the wired DSL providers, who sometimes give their modems to new customers for free, the dish is expensive. Data speeds are typically 1/2 Mbps downward, and 80,000 upward, which is comparatively slow.

Satellite Internet purveyors have a unique selling proposition. They can deliver wherever a clear shot to the sky exists, including farmlands, forests,

Figure 1-8
A satellite dish provides a 44,000-mile road to the global web. Photo by Advanced Network Services.

and islands that would typically be ignored by the other providers. The long-distance bounce introduces a lag between a request for data and the start of the data stream (called *latency*), which can interfere with interactive applications. They advertise a completion time of just two weeks after the order placement.

Fixed Wireless This Internet channel is technically similar to the Wi-Fi described earlier, except that the customer is expected to stay in one spot. Wireless DSL, sometimes known as *wDSL*, or *fixed-point DSL*, concentrates a low-power, high-frequency radio channel into a narrow beam in order to project it over distance. You will require a foot-square antenna (see Figure 1-9) on top of your house and an unobscured line of sight to a second antenna on an access point less than a few miles away. A professional will have to scale your roof with a signal meter in order to guarantee reception.

Figure 1-9 wDSL uses special radio antennas. Photo by Daly Road Graphics.

Unlike the previously mentioned pathways, wDSL usually works at the same speed coming and going, with a common data rate of 1.5 Mbps. Because most providers are also local ISPs, their Internet access is bundled with the data delivery service. This results in one-stop shopping, but limits your choices concerning the number of email names, web page space, and other service details.

Buying Internet Access

In the past, homeowners dealt only with a monopoly for their data needs. The telephone company gave customers little choice by today's standards. At one time, owning a telephone of any color besides black incurred an extra monthly fee, for example. But few choices also meant that few decisions had to be made.

Those days are certainly over, because customers must now wade though menu after menu of selections in order to reach reasoned decisions regarding their data supply. Therefore, you must plan your Internet access with the same care you take in planning the physical layout of your home network. You must first determine what you need and then balance that against what you can afford and what is available.

To begin with, if you are going to buy your data link from the phone company, as the largest percentage of users currently do, you must make a second choice as to who will provide your Internet service. That is a different matter, because in 1966 the FCC mandated telephone companies to allow for their own competition. Phone companies, or *Incumbent Local Exchange Carriers* (ILECs), can sell you Internet services, such as email and IP addresses. But you may also shop for the best deal from some other deregulated provider, such as AOL, Earthlink, Covad, or many others. For them, the FCC coined the term *Competitive Local Exchange Carrier* (CLEC). If your phone company cannot or will not provide DSL service, then one of these alternative providers might step in. Or you may have to wait.

As you read through the ads and take notes from ISP web pages, you should be skeptical. Remember that the word deregulated also means that you cannot complain to a government agency if you are unhappy with your purchase. Some large, well-established ISPs have gone broke and disappeared entirely, leaving customers high and dry. Your best assurance is to ask someone who already subscribes and to be fickle if your new provider's promises turn out to be bogus. Remember that fine print is made fine for a reason. Here are some of the questions you should ask before opening your checkbook:

- **How fast are the downstream and upstream speeds?** Claims regarding speed often refer to a "best case" or "best effort" and do not apply at peak usage times. Providers sometimes underestimate the number of users and the traffic they generate. The term for this behavior is the same one used by the airlines: *overbooking.*
- **Is usage unlimited?** Most ISPs let you surf all day, but if you eat up enormous bandwidth, they will notice. Some will make you pay extra for it. This can happen if you have several heavy users funneled into one connection with a router. Almost all ISPs forbid you to set up a web server on a residential account. They will either insist that you pay business rates or they will just say no. Most won't let you resell their service or give it away, as the previously mentioned freenets do.
- **Do you have to pay for the equipment?** Competition is so fierce that some providers will send you a DSL modem for free, just to sign you up. These promotions can come and go. It never hurts to ask.
- **Can you buy your own equipment?** Some ISPs will charge you an extra $5 a month forever to lease their modem, and take the modem back if you leave. Some will only allow you to use their particular

modem. Others permit you to buy a standardized cable modem from an electronic store. Telco DSL modems come in different varieties, so you may have to wait for your ISP to send one of theirs.

- **Can you hook the equipment up yourself?** The hookup price you are quoted often assumes that they will send equipment that you will install. They also assume that you already have some of what's required, such as NICs. Some ISPs, such as cable companies in metropolitan areas, don't mind dispatching people in vans to help you get started. Most will not help you install personal home networking equipment, such as routers, firewalls, and cabling. Some will shrug and use the slang term YOYO, which means *you're on your own*. It varies.
- **Is there a one-time startup fee?** As mentioned earlier, some ISPs will waive their initial connection fee, typically $100.
- **Do monthly charges go up after a while?** Make sure that the monthly fee quoted to you isn't part of a temporary promotion.
- **Does the contract allow you to add a router, and add multiple users on a home network?** At first, most ISPs did not want someone else's equipment connected to their modem. All insist that you remove it before they will troubleshoot an outage. But so many users are using routers, and it's so hard to detect when they do, that most ISPs (Verizon is one example) now advertise home networking as a value-added feature. Some will even sell you the equipment.
- **Do they provide backup dial-up service?** You cannot take your DSL modem on the road with you, at least not yet. So you will need a way to check your mail from a hotel room, and sooner or later your DSL line will go down. Then you will have to rely on your clunky old 56 Kbps telephone modem to keep your communications alive until the faster line is restored.
- **How many dial-up hours per month? Is it free?** If you are forced to use an ancient telephone modem because of a service provider's technical problem, will you have to pay for the backup? They may charge you from the first minute you dial in, and they may turn the service off if you use it too much. They may also refund the charges if you complain. It is best to know in advance.
- **How many email addresses will you get?** Your relatives, kids, and home business may all need their own email identities.
- **Can you read your mail if you are on the road?** Some ISPs provide browser-based access programs that enable you to check your personal email from anywhere. These will even enable you to do so from work, right through the company firewall that was set up to prevent it.

- **Will you get permanent or temporary IP addresses?** Most ISPs give you a temporary Internet address that changes from time to time. This makes it harder to do video conferencing or set up secure *virtual private networks* (VPNs) from home. Some ISPs charge more per month ($10 is typical) if you must have a permanent address. Some claim it's available except for certain areas that turn out to encompass several large states. Some cannot provide them period.
- **Do you get free storage space for a personal web page? How many *megabytes* (MB)?** Most providers are happy to give away 5 or 10 MB worth of disk space, because most customers don't use it all up, if they use any at all. You can use web pages to share family photos or sounds, advertise your resume, or spread news. The ISP will object if they catch you using a personal page for business purposes, or anything that attracts an unusually large number of viewers or is obscene or defamatory.
- **Do they throw in free software, such as a personal firewall, virus protection, and browsers?** Many sites on the web supply free software, but if you can get an up-to-date, certified firewall or antijunk mail program that their tech support will help you install, so much the better.
- **If you call their tech support line, how long do you have to wait?** DSL is reliable, but nothing lasts forever. Before you call their help desk, start a stopwatch. If you hear elevator music and a recording telling you to reset your DSL modem repeated more than a dozen times before a live human picks up the phone, then you should consider another ISP.
- **What's the soonest you can back out of the contract, and how much would that cost?** They won't like hearing that question. You may learn that your free modem and setup charges become billable if you ask for a divorce. They may have other fees ready to discourage you from leaving.

One place you can go for an independent assessment of prospective DSL suppliers is www.dslreports.com.

Connect Your Personal Computers

Wi-Fi can be useful even if you remain cut off from the Web. You may want to wirelessly connect your local PCs, laptops, and printers in the simplest way, which is called ad hoc mode. This is the Wi-Fi equivalent of wired *peer-to-peer* nets, in which all the components talk to each other directly without going through a central hub first. Many PCs and PDAs are equipped with infrared ports, which transmit data in the same fashion that your TV remote control does. These are fine for communicating from one to another, as long as they are in a line of sight, usually sitting right next to each other. But infrared ports cannot be used around corners or room to room, as is the case with Wi-Fi.

Each device on your Wi-Fi home network must be fitted with a hardware interface of some kind. This interface translates the data inside the PC, and reformats, transmits, and receives it among other Wi-Fi interfaces. You might say that they are the PC's walkie-talkies. For Wi-Fi, the data travels in the form of very high radio frequencies.

Avoiding Trouble Spots

The Wi-Fi range is limited and somewhat unpredictable because as radio frequency goes up, the wavelength becomes shorter and the signal behaves more and more like a light beam. That is, it will bounce rather than penetrate. That is the principle behind radar. Even decorations such as a potted fern will affect it.

A limited range is a blessing in disguise because it decreases the chances of interference from a nearby home or apartment's Wi-Fi. It also cuts the chances that someone may pick up your signal. (See Chapter 4, "Setup and Security.") How short is short? Industry specifications say anywhere from 75 to 300 yards, but it all depends on what stands between the receiver and transmitter. Radio waves at 2.5 billion cycles per second, or gigahertz, will penetrate thick plywood far more easily than thin aluminum foil, for instance. As a rule of thumb, Wi-Fi is sometimes referred to as a two-wall system, meaning that if three walls are in the way, you may be out of luck.

I conducted some simple tests on my own home system by walking around and watching the signal strength, signal quality, and link speed indicators displayed on a laptop screen. I found that if the access point was centrally located in the house, an adequate signal was cast all the way out to the edge of the property line. In the few locations where the signal strength displayed on the test laptop was low, reorienting (simply turning the laptop so it faced in another direction) restored an adequate signal. You will not be able to log in to a Wi-Fi system on the other side of town. If you put your access point in a front-facing window, however, you will probably be able to use your PC across the street. Plan your layout according to your needs.

Low signal strength doesn't equate to no data, just slower data. A Wi-Fi device will maintain a ragged channel by downshifting its transfer rate from a top speed of 11 Mbps to 5.5 and then 2 and 1, in stages. This process is known as *fallback*. At its slowest speed, the link is still 20 times faster than a dial-up modem. In our test, conducted from a home in southern California, we listened continuously to a netcast from a radio station in Cedar Rapids throughout the house, and lost the music only when we approached the front curbstone. Remember that a typical data rate over a DSL line is 1.5 Mbps. A chain is only is fast as its slowest link, and most of the time that link is going to be the one from the house to the web.

Some techies have extended their hot spots over multiple city blocks through the use of special antennas, either purchased legitimately or home built from Pringles cans or coffee cans (see Chapter 4, Notes on Homebrew Antennas). Don't do it yourself because it is not always legal. Incidentally, you may notice that many Wi-Fi devices, base stations in particular, have a pair of antennae. This design technique is known as *antenna diversity*, and its purpose is to eliminate dead spots in the coverage pattern that result from reflected or multipath interference.

Wi-Fi Interfaces

Wi-Fi interfaces come in different shapes and sizes, or *form factors*. Experts predict that someday soon they will be built into television sets, VCRs, and even automobiles. They are already built into standalone security cameras. When you shop, you should check for several characteristics. The item should be Wi-Fi certified, so that you aren't locked into buying from one manufacturer. Some cheaper cards on the market use the HomeRF standard, which is not Wi-Fi compatible. Some interfaces will operate on the newer 5 GHz frequency bands, but unless you plan to construct your network solely with that type of equipment, your purchase should also operate on the 2.5 GHz band. Some will handle both, but they cost more. If the device looks oddly inexpensive, make certain that it's not merely an adapter for a *Personal Computer Memory Card International Association* (PCMCIA) card, which will cost extra.

Personal Peripherals Many new laptops come with built-in Wi-Fi transmitter-receivers, just as they do with built-in telephone sockets, infrared outlets, and wired LAN cable sockets. So you may not need to add a PCMICA card at all. One example of a pre-equipped laptop is the upscale Toshiba Satellite Pro 6100, which has both Wi-Fi and Bluetooth interfaces built in at the factory. Older PDAs require an optional Wi-Fi adapter, but newer models have the hardware built in (see Figure 1-10). As PDAs

Figure 1-10 Hewlett-Packard's Wi-Fi-equipped iPAQ personal data assistant. Photo by Starbucks Coffee Co.

become more widely used for Internet access, more and more web pages and email servers are being redesigned to accommodate their simple displays.

PCMCIA Cards As seen in Figure 1-11, a PCMCIA is a compact circuit card in the approximate shape of a credit card. It fits into a corresponding slot on a laptop PC. In their most recent iteration, they are called *cardbus* slots. Such a card may also provide the driving hardware for larger devices by sliding into slots provided for them. Their service range may not be as far as other models, because these cards must have comparatively small, built-in antennas. Laptops were built to move around, and a longer antenna might get snagged or broken off. Some do have sockets for add-on antennas, which cost extra.

These cards can be easily removed and inserted elsewhere if the laptops are turned off first and you want to share the resource. As with all these interfaces, you will have to install driver software and enter some individual configuration information before your PC will know how to use them. Prices range from $75 to $140.

Peripheral Component Interface (PCI) Cards A *Peripheral Component Interface* (PCI) interface card is installed more or less permanently into a socket inside a desktop PC, much like an add-on sound card (see Figure 1-12). A few of these will also plug into the older *industry standard*

Figure 1-11
Cardbus Laptop Wi-Fi interface card. Photo by Starbucks Coffee Co.

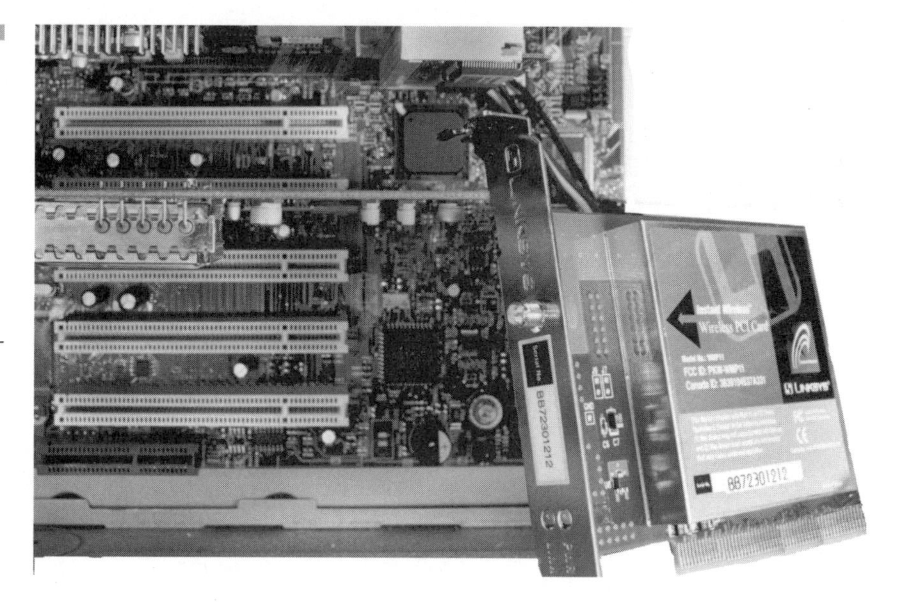

Figure 1-12
The four rectangular white sockets are motherboard PCI slots. The PCI-based Wi-Fi NIC, to the right, can be plugged into any of them. Photo by Daly Road Graphics.

architecture (ISA) slots. You do have alternatives if you are reluctant to open your PC and install hardware.

These are probably the least convenient interfaces to install, because you must have an empty slot available and then take your PC apart, observing antistatic measures as you go. Afterward, a short external antenna will attach to the card from outside the chassis. The PCI bus can move data in and out of a PC at 132 Mbps, which is more than enough to handle the fastest Wi-Fi transmission. Prices range from $40 to $140 dollars.

Universal Serial Bus (USB) A *Universal Serial Bus* (USB) interface is external to the computer (see Figure 1-13). It plugs into the USB socket, or port, which is now standard issue on desktops and laptops alike. These ports can be used to connect as many as 64 peripherals of many different types, such as cameras, external hard drives, and printers. The data on these small, special-purpose networks moves at 12 Mbps. USB devices, including Wi-Fi adapters, are designed to be completely plug and play, so that they can be detected and configured as soon as they are attached. They can also be unplugged and replugged elsewhere even while the PC is running to give maximum deployment flexibility.

One advantage of the separate USB interface is that you can move it around for best reception. Wi-Fi signals are very directional, and sometimes

Figure 1-13
An external USB Wi-Fi adapter. Inset: Close-up of a USB connector. Photo by Gigafast.

moving the base a few inches can improve the signal and throughput a great deal. The base can be semipermanently attached at a good spot with a small Velcro strip. If you want to share and move your connectivity, but not the computers, these are your best bet. Prices range from $95 to $140.

Wi-Fi Print Server One very desirable application for Wi-Fi has resulted in a special-purpose interface: the *printer driver*. As pictured in Figure 1-14, it is an external standalone box with a port for connecting a hard copy printer cable.

Using this interface, all the Wi-Fi-equipped PCs in your home can share that one printer, thus increasing the printer's usefulness and value. Some Wi-Fi access points include a printer port and have this function built in. Prices for standalones range from $130 to $330. Check to see how many parallel ports are provided and whether or not it will support Mac computers. Nowadays smart printers constantly report their progress, any alarms, and their ink supplies to the controlling PC, so make sure that your print server will handle bidirectional print functions.

Interfacing to your Existing LAN If it's Wi-Fi, then all these new components will work with the others. But what if you already have a wired network in place? Wired nets use a *hub* (see Figure 1-15), a device that

Figure 1-14
A Wi-Fi printer driver. Photo by Daly Road Graphics.

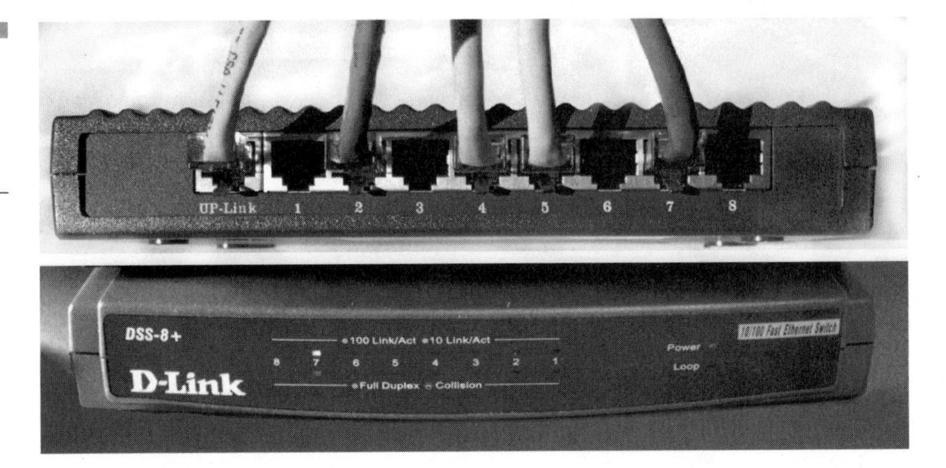

Figure 1-15 Front and rear views of a wired LAN hub. Photo by Daly Road Graphics.

interconnects all the wires and data from those wired computers. Rather than scrap your old net, it can be interfaced with a Wi-Fi net.

This specialized hardware interface is called an *access point.* Access points do not connect directly to a PC, but instead plug into your existing wired hub or into whatever modem you use to link your house to the Web. Access points are central, key components for most home Wi-Fi nets. They come in different levels of complexity and price depending on the tasks they are expected to do (see Figure 1-16).

The simplest of them presents one Ethernet cable to an existing wired hub, imitating a wired PC or a downlink from another wired hub. Every access point will have an antenna or two as well to gather input from your Wi-Fi devices. If many wireless devices are in use, or the distance between them is too great, then more access points can be added. They will also plug into the existing wired network to extend the coverage pattern. If a wire link between two separated locations does not exist, two access points can be set up to wirelessly *bridge* the two areas. As users move between multiple access points, they are handed off in a seamless manner similar to how a cell phone user in a car is automatically switched from one cell tower to another. Given a choice of which users to handle, the access points will communicate with each other and balance the traffic load between them. The theoretical maximum number of users on a single access point is 50 or more. But all active users share the 11 Mbps channel, so if many try to draw on it at once, each will only get a slow slice of the throughput pie.

Multifunction Access Points Most access points today combine their radio interface with a small, built-in wire hub of their own. They have a row

Figure 1-16
A typical, simple Wi-Fi access point. Photo by Belkin.

of empty LAN (called RJ-45) sockets, ready for wired computers to plug into. These ports will *autosense* the speed of the computer on the other end of the cable and appropriately switch between 10 Mbps or the fast Ethernet speed of 100. They will also interface two computers running at different speeds. Rather than simply echoing data from one PC onto all the ports, it selectively switches data from the source to target, leaving the other ports with capacity to spare.

Access points also have another 10-Mbps-only RJ-45 plug dedicated to connecting to the Internet interface, either a DSL or cable DSL modem. Some access points also provide the functions of a firewall and router (see Figure 1-17). Multiple-function access points perform the following tasks:

- Monitor traffic to and from the Internet to block out hackers.
- Selectively restrict outbound access.
- Log user activity and traffic in both directions.
- Assign IP addresses to the local client PCs when they boot.
- Selectively channel traffic from one home client to another, minimizing unwanted interference between clients.

Figure 1-17 An access point, a router, a wired hub, and a firewall in one box. Photo by Daly Road Graphics.

- Reject traffic from unwanted foreign systems.
- Encrypt data using WEP for greater intercept security.

You can set up or reconfigure these devices from anywhere on your net using your web browser. The industrial versions of these devices cost many thousands of dollars and are the size of microwave ovens. Home routers represent a remarkable amount of intelligence built into a small box for very little money.

Prices range widely from $130 to $500 dollars, because of the wide range of features that may be included. Some have built-in print servers, dial-out modems, or specialized interfaces for AC or telephone wiring-based networks.

Loading Wi-Fi Software

Congratulations! All the hardware has been unwrapped, plugged in, and turned on. Unfortunately, nothing will work until you load and configure the software brains so that the hardware muscle knows what to do. Chapter 4 will cover configuration in detail, particularly security issues. For now, we will present a summary.

Equipment manufacturers advertise their products as a breeze to configure. If you go along with the default values, or you know the values you want to enter, Wi-Fi devices are truly easier than they have been. Even so, as with business or industrial networks, every station on your home network will require some basic software modules. Specialized software known as *drivers* must be loaded into your PC so that it knows how to format data, direct it to the particular type of hardware interface in use, and then control the card's output. As with those PC cards, manufacturers provide very automated installation programs on CD-ROMs that speed installation.

Inevitably, you will have to make some additions and key in some unique identifiers of your own. Each Wi-Fi interface will need a unique Internet address, and a shared domain name and gateway address as well as the other parameters that you would set with a wired network interface. This is frequently confusing to new users, so much so that ISPs usually set aside one choice on their automated phone menus for that category of trouble call: "Push *X* if you've never gotten your network connection to work." They also provide tutorials and guides on their web page, which you should print out before you start at home.

Choosing factory defaults usually gets you on the air in a few moments' time automatically. However, all the experts quoted in Chapter 8 mention that additional work is necessary to gain security for your home network. In addition to common Internet parameters, Wi-Fi gives you the choice of encrypting transmitted data so that others cannot easily intercept it. You also must decide on the degree of encryption and the common key used to code data. The higher the level of encryption, the greater the throughput cost. Also you will choose which frequency channels the interfaces will broadcast on. All the devices on your network must match so that they can communicate with each other. Defaults will work unless you have a neighbor who occupied the default channel first. As more users buy Wi-Fi, it becomes more important to guarantee that a next-door neighbor will not inadvertently interfere with or log onto your home net, or vice versa.

Wi-Fi access points and routers are configured and controlled using a built-in web browser interface. After you log in to one, your access point parameters are displayed as though you would connect to an Internet web page. Because they interface with the ISP, the DSL modem, and all the other devices on your home network, you will have to answer the following questions:

- What is the type of connection to the ISP? (The *Point-to-Point Protocol over Ethernet* [PPPOE] requires an automated login, like telephone modems, and needs your password also.)

- Will it have a static network address of its own or fetch one from the ISP?
- Will your router supply addresses to your local Wi-Fi stations?
- How many stations will there be on your network?
- Will some stations, such as web servers, be accessible from the Internet?
- Is data being encrypted? How heavily? What is the password?
- What is the name of this home network? Do you want to advertise the name?
- Do you want to allow members of a foreign network to join yours?
- Will you want to track the comings and goings of the users?
- Which channel is the radio interface set to?

Some access points can even use virus software to check every incoming data stream. After the other features are working, you must pay for and download the virus software, and then load it into the access point. Of course, you must check for and download virus updates. For that matter, you should check to see if any of your equipment manufacturers have patches for any of your Wi-Fi devices. Lastly, when your configurations are tweaked to perfection, back them up so they don't get lost. Some APs include a facility for uploading configurations to a PC hard drive, but the easiest way to save them is to write them down.

Ready for Takeoff

You will know your network is finished when you don't notice it anymore and instead devote all your attention to the data riding on it. After turn-up day, you will notice some changes, but you will quickly come to take them for granted.

For example, your PC will wait all day for messages and announce them as soon as they arrive. Other members of your family will work or play with the World Wide Web, but you won't notice that from your terminal; it will seem as though you are the only one logged on. You will stop copying material from the Web to store on your hard drive, because you can get it from the Web whenever you need it just as quickly. The huge files that you used to dread, such as Movie Maker films, will just make you chuckle. You will haul your laptop around the house as you do your cordless phone (see Figure 1-18). Your kids will still nag you for a faster PC though. Some things never change.

Figure 1-18 Work from the poolside as you would from the office. Photo by Daly Road Graphics.

Hopefully, this first general review chapter has answered the question "What is Wi-Fi all about?" At this point you should know whether or not you want to investigate further. If so, then you probably have more detailed questions about the issues you have discovered here. We will dig into the many details and expand and define Wi-Fi for you in the chapters to come. We will resume in Chapter 2 by taking a closer look at the problem of connecting your house to the rest of the world.

CHAPTER 2

From the World to Your Door

This chapter lists the prerequisites for connecting a home system to the Web. It explains in greater detail how a *digital subscriber line* (DSL) works, and its costs and options. Descriptions of Internet addressing and web servers are provided. The pros and cons of buying from the phone company or an independent *Internet service provider* (ISP) are discussed. Service (bandwidth) levels and the commonly offered bundled services are listed with an analysis of their usefulness. Also covered are frequently asked questions, such as how long it takes to get or install bandwidth.

Into the Fast Lane

Although a home *wireless fidelity* (Wi-Fi) network can easily interconnect home-based PCs, the main attraction of having a PC is the wealth of information just a few clicks away on the Internet. News, weather, sports, email, research, long-distance gaming, business—these are only a few of the reasons to want your home net to be an extension of the World Wide Web. After a high-capacity broadband line is completed to your home, you may distribute data from it to several PCs. (We will describe that process in Chapter 3, "From the Doorstep to the Tabletop.")

According to researchers at the Pew Internet and American Life Project, users who forsake slow telephone modems for broadband access

- **Create more content for other web users.** On a typical day, 16 percent of broadband users update their web site or post to web diaries or chat rooms, for example. Only 3 percent of dial-up users create content on the average day.
- **Learn more.** Eighty-six percent of broadband users say that the Internet has improved their ability to learn new things, as compared with 73 percent of dial-up users.
- **Go online more often than dial-up users.** Eighty-two percent of broadband users are online on a given day, compared to 58 percent of dial-up users.
- **Shop online.** Thirty-one percent of broadband users say the Internet has reduced the amount of time they spend shopping in stores.
- **Work from home.** One-third of broadband users telecommute occasionally. Fifty-eight percent of broadband users who telecommute say they spend more time working at home because of the Internet.

- **Surf more.** On a typical day, a broadband user does about seven online activities (such as news, healthcare, or hobby surfing). By contrast, dial-up users do three Internet activities on the average day.
- **Use the Internet to study.** When asked about their most recent major school report, 71 percent of teenagers with Internet access said they relied upon Internet sources the most in completing the project. That compares to 24 percent who said they relied on library sources the most (according to Pew Internet Project's Broadband report).

DSL, from whatever source, is billed at a fixed price monthly, most often without regard to actual usage. Exceptions occur, which we will explain later, but usually your bill will be the same whether you just pick up your email or "surf 'till you drop." After you have your broadband line, you can use it any time night or day, and see any or all parts of the Internet. You won't have to log on because even if your ISP requires a username and password, they will be supplied automatically.

No modern business would expect their employees to function effectively without Internet access. Naturally, those of us who work from home can be more productive with a dedicated *wide area network* (WAN) line. It follows that homeowners and apartment dwellers can benefit from the knowledge and communication available on the Web as well. Broadband is more widely available to residential users and less expensive than ever before.

How to Connect a Broadband Pipe

Presently, four methods are available for bringing the Internet to your doorstep. Your choice will depend first on their availability and then, if you have more than one choice, on price:

- *Asymmetric digital subscriber line* (ADSL), which uses a high-frequency signal impressed onto your telephone line
- *Cable DSL*, riding unused channel space on your cable TV line
- *Satellite DSL*, which uses the same link as satellite TV, only bidirectionally
- *Wireless DSL* (wDSL), which is long-distance Wi-Fi

All the previous methods have some characteristics in common, and regardless of the media used to transport it, DSL is sometimes referred to as ADSL.

How Fast Is Fast?

Asymmetric, or unbalanced, refers to the common practice of having two transfer rates: a fast one (typically 768 Kbps to 1.5 Mbps) for information flow *into* your net (the *downlink*) and a slower *uplink* speed (384 Kbps) for data *outbound* to the net. Residential ISPs operate that way regardless of the transmission media because it's cheaper, and most home users only transmit when they are requesting data. Those requests are brief and usually come through the computer keyboard. Since most of us cannot type at 187,000 characters per second, a slower line speed for that function is more than adequate. Perhaps the only time you will notice the difference is when you forward a mail message with a long file attached. It will take longer to send than it took to receive. A home-based business operator (someone with a web server in his or her back room, for example) may want to send information to customers with a wide-open throttle. A *symmetrical DSL* (SDSL) link is available to them—for a price, of course.

The speeds we will quote represent the best case. Most ISPs cover themselves by promising a best effort to achieve them. A sad fact about networks generally, including Wi-Fi, DSL, and ISPs alike, is that providers often measure throughput with a rubber speedometer. At least 13 percent of their stated maximum will be lost through overhead costs. Individual circumstances do vary. If the uplink is saturated to capacity by sending a large file, then the downlink can drop to almost the same speed. That's due to a built-in deficiency in the *Internet Protocol* (IP), and all ISPs are subject to it. Cable Internet is more vulnerable because all the users on a single cable segment must share the uplink. The *Data over Cable Service Interface Specification* (DOCSIS) recommended maximum is 1,000 users per segment. It's not hard to see why ISPs are intolerant of users who provide some kind of nonstop streaming service, such as webcasting MP3s on a residential line. It only takes a few so-called abusers to set off a traffic jam.

Whether your data is delivered over cable television or telephone lines, your DSL provider will not necessarily be the same as your ISP. If you have an email address that all your friends are familiar with, you may want to keep your dial-up ISP and simply upgrade the service to DSL. If you do, most ISPs will let you keep on using your dial-up for a few hours a month (20, typically) either as an emergency backup for the DSL link or as a way of checking your email from a hotel room. They'll let you use it as much as you like usually, if you are willing to pay for the extra service.

Unlike *plain old telephone service* (POTS), DSL *quality of service* (QoS) is not regulated by a state agency. If your link is giving you grief, you cannot call the *Public Utilities Commission* (PUC) to complain about it. Actually, thousands and thousands of irate customers do so anyway every year to no avail. ISP dedication to customer satisfaction can vary unpredictably. Providers are caught in a perennial squeeze between falling revenues, intense competition, and the urge to claim as many new customers as possible, for new customers are easier to acquire than someone else's. Users cannot assume that their provider will be as permanent or as well equipped as the telephone companies they are familiar with. For that matter, after the bankruptcy of WorldCom and newsreel footage of Adelphia executives being led off in handcuffs, users cannot assume the permanence of their phone companies or ISPs. Most providers will insist that new customers commit for an extended period of time, in order to qualify for signup incentives. But it was only in 2002 that ISPs were required to warn their customers at all when the provider decided to go out of business. The restriction was a legislative response to service providers that quit during the dot-com collapse of the late 1990s, such as North Point Communications of San Francisco, which suddenly abandoned users when going out of business. In such a circumstance, it actually takes longer for a user to transfer to a new provider than for the initial signup and installation.

You can go to places on the Web in advance of signing a DSL contract to get independent estimates of their service and track record. One such is www.dslreports.com. They provide several services, including

- Helpful installation and troubleshooting tips
- News items about the cable and DSL industries, and government regulation
- Service locators for your area
- Frequently asked questions
- Tweaks and adjustments to boost speed
- Forums and opinion polls
- An automated security check of your system

It costs you nothing to look, and when you see their criteria and a few sample track records, you will form your own questions to ask. It also costs you nothing to listen. As the old saying goes, "Ask the person that owns one." Your neighbors or coworkers may have tales to tell you.

You should also remember that any prices we quote here would vary from week to week, as vendors compete and special offers come and go. They are intended only as a general guide. You will have to check vendors' web pages individually to be sure. With prices generally declining, you may be pleasantly surprised.

If broadband Internet Service is as fast as everyone claims, you may be wondering if your old 486-type microprocessor can keep up with it. The short answer is yes. The long answer is that it all depends on what you plan to use the Internet to do.

If you plan on interactive 3-D games, your old clunker may fall behind, but the problem will be the PC's video card or processor against the complexity of the application, not an excess of data spilling over. For most applications such as web browsing and email, an older machine will do just fine.

Your Internet Service Provider (ISP)

Regardless of the media you decide upon, your ISP will be your true entrance point to the Internet. This may or may not be the same firm that provides the medium. When you send IP packets from your home network to the world, they aren't really on the Internet until they get to your ISP, who then channels them onto the Internet through their gateways. Your ISP does other necessary things, such as giving your home network an IP address, either static or temporary. It provides facilities to translate names to IP addresses, called *domain name servers* (DNSs), so you can ask for a web site by name and still reach it by its IP address. They provide other services too, such as email and personal web pages. You may not need all the services they provide, even though they are bundled with the ones you do. Usually, the ISP is the agency you call first if you need your bill straightened out or some kind of technical help.

The phone companies that own the lines sell Internet access too, and they would dearly love for you to buy it from them. But no law says you have to. In fact, the Telecommunications Act of 1996 says that telephone companies have to provide for their own competition by allowing you to shop for the best ISP deal. The dual-personality Telco/ISP arrangement can complicate troubleshooting a dead connection, because each half of the team can examine only part of the puzzle. This sometimes tempts them into jealous finger-pointing contests, with delays every time the problem is handed off between the Telco and the ISP. Both want a happy customer, but neither is motivated to make their competitor look good. A handoff lag can also be a problem between departments within either agency. (See Chapter 6, "Can You Hear Me Now?")

Read the Fine Print

However you connect to the Internet, you will be trading your money and promises for a provider's bandwidth and promises in return. It pays to take a few moments to compare offers, shop, and, if necessary, walk away from a lame deal.

Here are some questions to ask and some catch-22s to avoid. First, realize that the initial signup giveaways are predicated on the expectation that you will be a loyal customer for years. If you are fickle and pick a competitor later, you may discover that you've agreed to pay back all those discounts. You may owe them a refund even if you have to move out of town because your company requires it. Of course, if you don't mind paying more, no-contract service is almost certainly available if you ask for it, and you may have to pay for installation and startup. Subscribers with contracts are worth more to an ISP, especially the ones that might be considering a merger or sellout.

If you can avoid the aforementioned penalties, changing to a new ISP is usually easy, assuming that the replacement uses the same DSL carrier. If not, it works as though you have completely shut your connection down and are rebuilding it from the beginning.

The Frequently Asked Questions we cited in the first chapter apply to almost all prospective customers. Here are some other gotchas that are less general, but still worth considering.

- **Are their Fair Use prohibitions going to cramp your style?** Most providers have a list of prohibitions including obvious no-nos such as spamming the world, publishing pornography, libeling an organization, or preaching violent hatred. These probably won't apply to you, but the provider's list can be so long or vague that they can dump you without an appeal for doing something you consider legitimate. If you are buying a high-speed link to solve a specific problem, such as a web site for my kids soccer team statistics, ask them about it.
- **Does Fair Use include forced traffic reduction?** Some providers, particularly satellite DSL providers, will automatically strangle your speed if you exceed their usage limit. The rules defining their trigger event and their throttle duration may be complex. In practice, they may also prove to be very inconvenient.
- **If they supply home networking equipment, then who is responsible for fixing or replacing it?** If you are just going to be handed off to the manufacturer anyway, then you are probably better off shopping for it on your own.
- **What will the routing be from your home to the Internet?** If you've signed up with a national provider, your packets may have to cross state lines before they actually hit the Internet. That won't decrease your speed, necessarily, but it will affect latency, which is very

annoying to Internet game-players. And it may complicate troubleshooting.

- **If you plan to use your own built-in DSL or network interface, will it be compatible with their equipment and protocols?** Some PCs have interface equipment built in. You may want to use a cable modem you carried over from your previous residence. They may not work now.
- **Will additional interior or exterior wiring be necessary?** Usually, the answer is no, particularly for a Wi-Fi home network. But if your broadband link must terminate in a designated location, such as a computer room, you may have to pay someone to put it there.
- **Are their restrictions for my particular PC or operating system?** Pentium or clone-based PCs are universally accepted. But will the provider's installation CD recognize a MAC, or a Linux, or a Unix, or something that's not quite so standard?
- **How long does it take for service to be activated?** Rest assured, it will take longer than ordering a pizza. Whatever answer you do get will be a theoretical best case.

Your Installation Kit Satellite DSL modems most often connect to your PC over a Universal Serial Bus (USB) interface, but the self-installation kits used to connect your PC to other types of DSL modems may be a variety of other form factors. Each has advantages and disadvantages. Providers favor the USB types because they are simpler than the more common method, which consists of a short Ethernet jumper cable between the modem and a network adapter inside your PC (see Figure 2-1). Simpler equates to less user frustration because installing and configuring IP settings for an internal PC card is beyond the abilities of some first time users. USB results in fewer trouble calls to the service provider. A second advantage, from providers point of view, is that USB-type modems make it more difficult for you to share a single connection among multiple computers. (DSL routers use the Ethernet cable interface.)

Some providers may send you a DSL modem on a card that installs inside your PC, which is the worst of both worlds. You'll have to take your PC apart to install it, as with the network cards, and you will be out of luck if you want to buy a router to service several PCs. If you have only one PC in your house and you don't like to fiddle with technical stuff, then USB is best for you. If not, you can always tell your provider that you don't have any open *Peripheral Component Interface* (PCI) slots left and request a modem that uses an RJ-45-compatible interface. To save yourself a delay

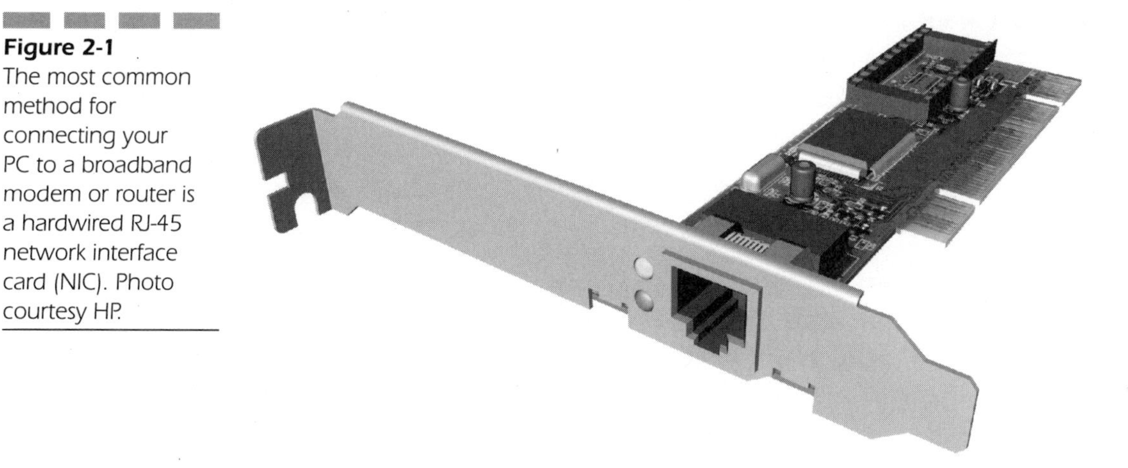

Figure 2-1 The most common method for connecting your PC to a broadband modem or router is a hardwired RJ-45 network interface card (NIC). Photo courtesy HP.

and hurt feelings, you should ask your new provider in advance what kind of kit they plan to send to you.

If you have a newer PC, it may have a DSL modem installed on the motherboard at the factory. Unfortunately, not all the DSL providers are standardized to the extent that Wi-Fi manufacturers have. Your built-in modem may or may not work. You'll have to call and ask.

The High Road: Satellite DSL One Internet connection path that's almost always available to you is the one in the sky. Satellite television services offered by DirecPC and Starband offer Internet access through their high-orbit outposts via the same technology that gas stations have been using to report your purchases when you insert a credit card at the pump. (You may have noticed that most filling stations sport a meter-wide, oval, skyward-pointing dish these days.) That's the same kind of expensive parabolic antenna that you will need on your roof, which counts as a disadvantage. In the past, satellite providers relied heavily on the asymmetric service scheme, because users actually used earthbound telephone lines to uplink their requests to the Internet. Only the downlink data came literally down from space. Some users still do it this way, though providers are prodding them to upgrade to a genuine two-way link by citing many of the disadvantages that former telephone modem users are familiar with.

Because two-way dishes transmit as well as receive, a professional must install and precisely aim that dish (see Figure 2-2). The installation process will take about an hour and a half. Part of that time will be spent removing your existing satellite dish and running a second (data) cable to the new one. Naturally, you will have to install the provider's driver software. Your

Figure 2-2
A close-up of a satellite Internet dish's feeder horn. Photo by Advanced Network Systems.

PC will have to be configured as a software gateway if other networked PCs are to concurrently share the satellite link (See Chapter 4, Connection Sharing). Gateway computers must be left on continually for the clients to stay online.

The same dish can be used for the pickup of video and data; however, the transmitting satellites are separated by a few degrees in the equatorial Clarke Belt, so focus will be fuzzy on the video side. This usually won't make a difference except in rainy weather. Otherwise, your antenna will require an unobstructed view of the southern sky. The 44,000-mile round-trip distance for the bouncing signal introduces speed-of-light drag, resulting in a noticeable (half-second) pause between a request for a feed and the start of the stream. This latency makes a satellite path too clumsy for tasks requiring immediate feedback, such as multiplayer video games.

Satellite data speed is often slower than land-based links. Providers advertise 500 Kbps for the downstream link and 80 Kbps for the upstream link. Again, this is a theoretical best case. It can be slower when rain fade degrades the signal or when you do not have a completely clear line of sight to the satellite. This happens often enough to merit the installation of a dial-up backup if you are running a full-time application. It will launch automatically if contact with the satellite is lost. Your screen will display a message when contact is restored, so you can manually shut down the phone link.

Your service provider may subject users to what is euphemistically termed a *fair use policy*. Such a policy states that if you download more than 170MB of data in a 1- to 4-hour time period, the company might strangle your bandwidth to slow speeds for another 8 to 12 hours. During off-peak hours, defined as 2:00 A.M. to 5:00 A.M., users can download 225MB of data. If you think you need more guaranteed throughput, you can buy more, but it will cost more.

Providers advertise a two-week interval between order and turn-up. Internet service may either be bundled to include 10 email accounts and 10MB of web page host space. You may be allowed to keep your present ISP, depending on which satellite company you choose. Their ISP cost is $60 per month with a one-year commitment. Equipment and other initial setup costs (typically $500) make it more expensive than your other choices, if you have any. That's pricey, but if you live in the Yukon Territory, it will get you on the Web, and you can also use it to watch TV during those long polar nights.

At least one of the satellite TV companies is selling DSL service that actually comes over a phone line, just like the phone companies. That may sound confusing at first, but it has nothing to do with your satellite dish.

Cable TV DSL Your choice between the other pathways is likewise determined by which ones are available in your area. *DSL over cable* is pretty much the same as DSL over a phone line in terms of speed. Cable may even be a bit faster if their system is new or well maintained. But the carrier signal is *radio frequency* (RF) and subject to interference. The infrastructure has to be re-tuned periodically. Cable companies are tempted to put off this maintenance to save money at the expense of signal quality. This can affect cable data more than video. Cable providers don't give ironclad guarantees concerning consistency of speed or latency.

Cable TV companies are in the "land rush" to sign up new customers, but few of them have completely overhauled their infrastructure to accommodate two-way data. This requires a cable modem on the customer end and a *cable modem termination system* (CMTS) at the cable provider's end, plus a lot of equipment in between.

Because of its great bandwidth, coaxial cable can carry many TV channels, each getting a 6 MHz slice of the total available. Internet data is encoded to look just like one of those TV channels, and occupies the same bandwidth. As with telephone DSL, cable upstream data links have a slower data rate. They fit into a smaller 2 Mhz frequency slice. Unfortunately, these uplinks ride on the low end of the cable frequency spectrum and are therefore more subject to interference.

Most systems are upgrading to fiber-optic cables. Their bandwidth is even more enormous. They carry digitized TV and data channels far out to the neighborhoods without distortion, and there it is then broken out onto coaxial lines for the short hop into residences. You may be able to get Internet service even if you are not a subscriber to the cable video, but if so, it will cost more than if you were getting the TV channels.

A typical monthly fee is $40 to $50, but you may get a discount if you subscribe to cable TV service at the same time or if you buy your own modem (which will cost you from $100 to $300). One advantage over the other pathways is that cable DSL is usually faster to get going, assuming that you already have a cable TV line into your house. The hardware requirements are simpler because the cable modem will probably adhere to the DOCSIS standard. If so, the modem can be purchased off the shelf (see Figure 2-3). This is almost always a better deal than leasing one from the cable company, because if you do, you will be charged month after month, regardless of how often you actually use it. Buying a non-DOCSIS or proprietary cable modem is a bad idea, assuming you can find one that matches your system's specifications. They can become obsolete and it may be hard to find someone to repair them. Before buying a proprietary cable modem, ask the cable company if the device will remain useable in the future. The cable franchise may be planning infrastructure upgrades that will render it obsolete. And ask yourself if you plan to be staying in your present area, because the operator in the next town may use something else.

A common complaint about Internet service in general and cable DSL in particular is that providers sometimes underestimate the demand for service and fail to provide adequately for it. The term for this sometimes pre-

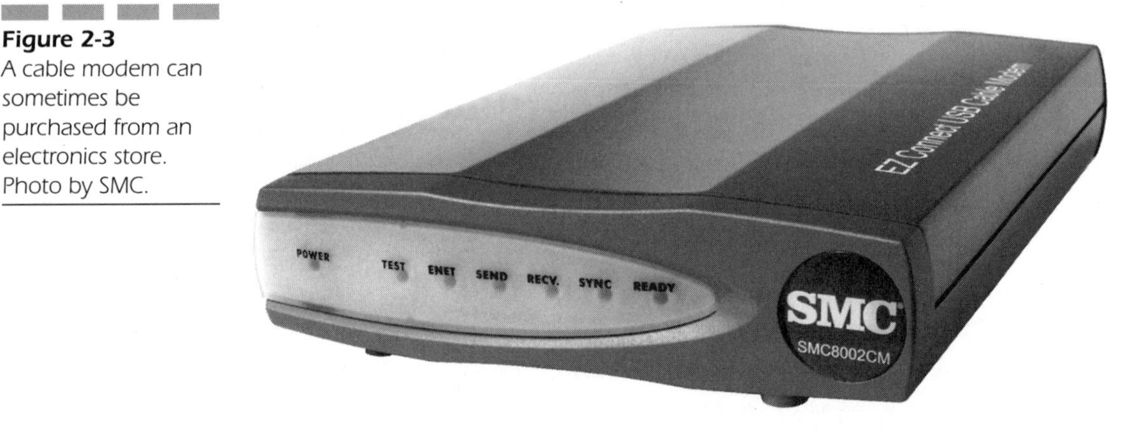

Figure 2-3
A cable modem can sometimes be purchased from an electronics store. Photo by SMC.

meditated practice is the same one used at the airlines: *overbooking*. If too many subscribers are riding the same cable, the service may bog down. Users may have to avoid cable Internet rush hours, such as Sunday evenings and weekdays at 6 P.M. when users arrive home from work and check their email. Users who play interactive games or do video conferencing may get frustrated at those times.

You may not have to purchase cable TV service and cable Internet service as a bundled package, but if you do so, you can watch TV and surf the Net at the same time. If you are already a cable TV subscriber, your existing cable may have to be replaced because the cable modem needs a higher-quality signal than your TV set. Also, if you are using a splitter or amplifier to drive multiple television sets, make sure the cable modem is connected ahead of it, because multitap devices will not pass bidirectional signals. With telephone DSL, an unshared telephone line carries your data back to a central office. But a cable line is shared with many other users, which makes security measures for your home network more important. When you share your PC's hard drives one to another, be sure that you are not sharing them with the world as well.

To summarize, cable and Telco DSL are usually equivalent in terms of price and performance. Telco DSL promises to function more quickly in the future, however, and offers an exclusive upstream link to the Internet. If you hope to upgrade to an SDSL business-grade service, Telco DSL gives you a path.

Ma Bell DSL

Spurred by relentless competition, the telephone companies are anxious to sign up as many customers as they can. Virtually every major telephone company in Europe, the United States, Canada, and throughout Asia-Pacific has announced commercial DSL rollouts. According to technology market analysts *International Data Corporation* (IDC) in its report published on February 9, 2001, "The number of digital subscriber line (DSL) subscribers around the world will soar to 64 million in 2004." This wide availability means that many users who want to get DSL service can get it from more than one supplier. They can choose to take service from their telephone company and often directly from their ISP or even work with multiple ISPs that best meet their needs.

Telco DSL has a security advantage over cable or fixed wireless in that it uses a private, point-to-point connection to the telephone company central office. Nearby customers do not share the line so they have no access to other user's data. It also means no shared bandwidth constrictions.

How Telco DSL Works The telephone company's digital subscriber service uses a signal impressed on your telephone, sometimes referred to as a POTS line. In principle, it works like the old telephone modems, but phone modem throughput was limited because it used the audio frequency range (500 to 2,500 Hz), which the human ear can detect and the phone system was originally designed to carry. By raising the frequency of the carrier tone far above the range of human hearing (4,000 Hz), the phone company engineers found that they could make it carry much more data and do so without interfering with the lower frequencies that we listen to. The scheme also gave the added benefit that people could continue to use the phone as they always had—for talking—at the same time that the line was carrying data. That's one of the reasons why DSL is the world's most popular method of connecting to the Internet (see Table 2-1).

The most common DSL offering is designated ADSL Lite, or G.lite, which is an *International Telecommunications Union* (ITU)-sanctioned standard. It offers download speeds up to 1.5 Mbps and uploads at rates up to 384 Kbps. Full-rate ADSL is faster, offering a download rate of up to 8 Mbps and an uplink rate of 1 Mpbs. But the full-rate standard is more than most residential users will need, and it requires that phone company installers install a special splitter on your phone line to separate the voice from data. It also costs a lot more. One advantage of Telco DSL is that the medium is

Table 2-1

23.3 million users make DSL the most frequently used method of Internet access. (Cable users total 15.8 million.) Table courtesy of the DSL forum.

Region	Total DSL Subscribers	Residential DSL Subscribers	Residential % of Users
Asia-Pacific	7,949,000	6,970,000	87.7
North America	5,510,000	4,267,000	77.4
Western Europe	4,232,000	3,523,000	83.2
South and Southeast Asia	499,000	374,000	75.0
Latin America	380,000	271,000	71.3
Eastern Europe	53,000	32,000	60.4
Middle East and Africa	48,000	37,000	77.0
Global Totals	18,671,000	15,473,000	82.9

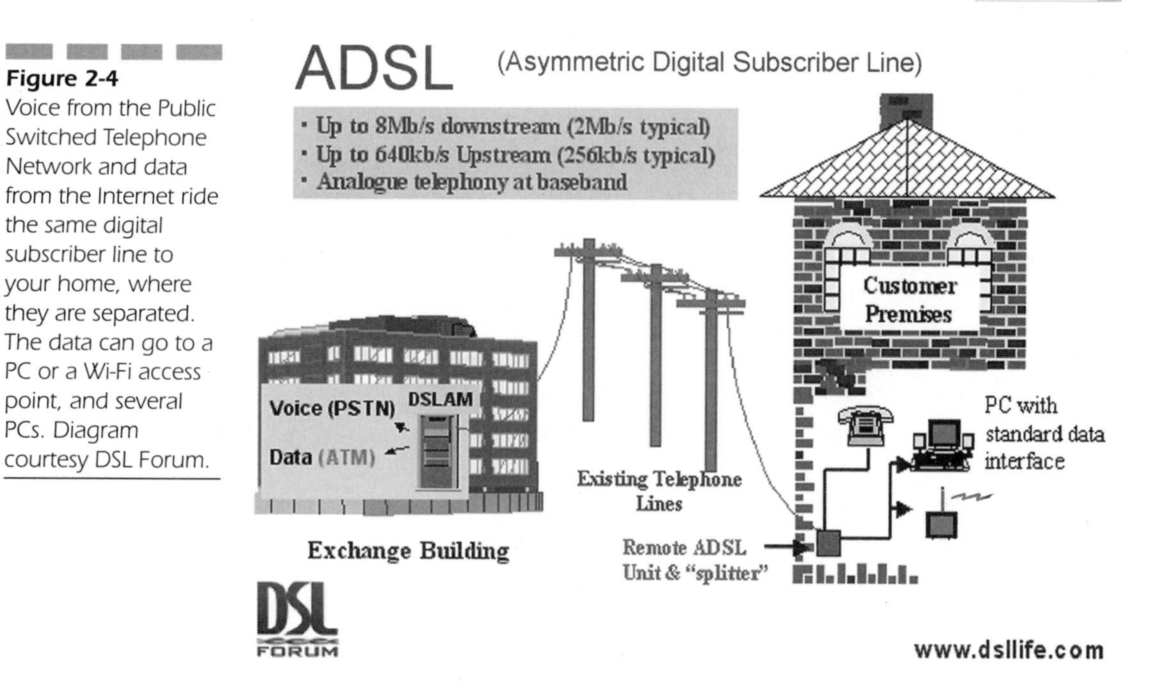

Figure 2-4 Voice from the Public Switched Telephone Network and data from the Internet ride the same digital subscriber line to your home, where they are separated. The data can go to a PC or a Wi-Fi access point, and several PCs. Diagram courtesy DSL Forum.

technically ready to go even faster than 8 Mbps. With refinements, future throughput might improve to 12 Mbps. See Figure 2-4.

Telco Distance Limits

DSL service is limited to those lucky customers who happen to live within three miles of a *central switching office* (CO). The throughput rate will vary from ISP to ISP. The speed available to users is limited by line quality and distance. The typical 1.5 Mbps residential offering assumes a maximum length of 12-thousand to 17,500 feet. The range limit for higher or symmetrical rates is less than that. Those near the end of the line may have to settle for less than top speed in any case. Some lines, though short enough and adequate for voice, may not be up to snuff for high-speed data. Even with a perfect new telephone line, the digital signal degrades over distance. The high carrier frequencies deteriorate even more quickly when forced through *loading coils* that are installed to condition long lines solely for voice. *Bridge taps* are unconnected cables that remain spliced onto your telephone line, probably as leftovers from an old connection to a different subscriber. They are notorious for soaking up data frequencies. If your home line attaches to a CO through a digital *pairgain* link, you cannot get DSL at all.

Customers who have waited all day for a telephone repairperson to show up might worry about getting DSL service from the same source. Usually, installers do most of their work outside the house and in the background. If they have access to the terminal block on the outside of your home, they can complete the final check without having you present. If you should buy the DSL service from another entity, it is still the telephone company that runs a new copper line for the service up to your building, or to the *demarcation point*. Your DSL provider must then show up to do any inside work. Sometimes the two don't communicate well under this arrangement.

ISP employees used to routinely show up and personally install DSL hardware and software at customer sites, but they learned that most customers were willing and capable of doing it themselves. Customers also prefer to save the service charges and forego delays.

Delivery Times The technical uncertainties concerning the DSL signal sometimes leave the phone company with a choice of either cleaning up your telephone line or turning you down as technically not feasible. If they decide to overhaul the line in a process known as *conditioning* (they are not required to do so) and additional problems occur between the phone company and your chosen ISP, a couple of months may pass before you get a *firm order commitment* (FOC). This is the date at which outside wiring is scheduled to connect to your *minimum point of entry* (MPOE). Your ISP should let you know when that is and whether or not you will have to be present to let the telephone installers into your yard. Depending on how the trunk lines are distributed, you may be eligible and your neighbor may be out of bounds. Fortunately, the phone company can send a test signal down the line before you buy. This evaluation will usually, not always, prevent a disappointment when the delivery truck drops off your new DSL modem. Over time, the Telcos and ISPs are becoming better coordinated. Some of them can team up to deliver service in a week.

Different varieties of DSL modems are available, and the kind that connects to your telephone line will depend on the equipment your phone company uses at the other end of the line. If you do not select your phone company as your ISP, you may wait for your ISP to deliver the appropriate modem to you. That can happen weeks after your telephone company says the line is ready. Even if you live across the street from a switching office there may not be any DSL equipment inside of it, ready and waiting for your connection. Telephone companies are installing fiber, routers, and *Digital Subscriber Line Access Multiplexers* (DSLAMs) as fast as they can, or rather, as fast as they can afford to. But in order to bill the most new users first, they prefer to install in concentrated urban areas, so that a CO may be fully utilized from the day it goes online. Smaller companies, known as *Competitive Local Exchange Carriers* (CLECs), have jumped into the race to grab and hold new customers, but if you live in a rural area, you still may have to wait until someone gets around to you.

Most ISPs provide a range of horsepower and service choices, bundling in extras such as multiple email addresses and personal web pages. The competition for new customers is so great that many are giving away the initial setup costs, the DSL modem, and throwing in a cut monthly rate as signup incentives. Many actually lose money on the deal at first, hoping to gain profits if you stay happy and stay subscribed for years. Typically, $70 per month will get you 1.5 Mbps downstream. That price gets hiked if you declare yourself to be a business customer. Shopping around will save you far more than it will cost. Although you may have trouble getting through on the tech support line, you will find that the salespeople will pick up their phones comparatively quickly.

Fixed Wireless The last of the four linking methods is actually another version of home Wi-Fi technology. With fixed wireless or point-to-point DSL, a high-frequency radio signal replaces ground cables, but the signal is focused on a line of sight between two directional antennas. To connect in this way you will need an antenna on the roof of your house or business that measures roughly one foot high and two feet wide. From this antenna you must have an unobscured line of sight to a fixed wireless access point that is within range. Alignment is not as critical as with satellite antennas, but it will still have to be done by a servicepeson with special test equipment. Somehow someone will have to run wires from the roof down into a central access point in your house that provides a standard RJ-45 cable link to your internal network. You will have to distribute connectivity from there, perhaps with a hub or Wi-Fi access point.

Latency is negligible, equal to wired networks, and overall performance is usually good as well, with a typical service target of 1.5 Mbps for the downstream and 128 Kbps for the upstream. As with cable DSL, wDSL subscribers must share the capacity of a given access point, unlike Telco DSL, in which each subscriber has his or her own line back to a CO. It is possible for wireless to get slow during periods of heavy traffic, especially if the channel has been overbooked.

AT&T's version includes *voice over IP* (VoIP) telephone service for $80 per month. Sprint gives you data only for prices equivalent to wired DSL. As with satellites, fixed wireless is handy for connecting businesses or communities that do not have alternatives. Some have started as grassroots endeavors by individuals in data-starved communities who have banded together in order to set up and share connectivity to the Internet. They share the wireless pipeline to an out-of-town access point.

Installing Your DSL

One of the corollaries to Murphy's Law states that "Before you do anything, you have to do something else first." All the equipment and service providers have attempted to make the installation process as simple as possible, but it can still be confusing. A little preplanning will make your installation much less frustrating. You may receive your boxes of hardware before the turn-up date—that is, the due date for the activation of DSL service. If so, you can use the time to get squared away on your end of the connection.

Your first step, after ordering the service, will be to determine where you want the equipment to reside. In the simplest case, you will have only one PC, so a Telco DSL modem will be installed in the phone outlet nearest the PC, perhaps the same outlet you used for your dial-up modem, if you are upgrading from one of those. If you are going to use cable DSL, the modem will have to connect to your cable outlet, which is typically close to your TV set. If your computer room is somewhere else, you will either have to run a LAN cable to it or invest in a Wi-Fi interface to spare yourself that hassle. In either case, you will need a full-time electrical outlet to power the modem, not one that gets turned on and off frequently. As we mention in Chapter 5, "Good Housekeeping," some kind of surge suppressor for all your computer equipment is a wise investment, if for no other reason than to provide you with the multiple outlets you are going to need to power the modem, the PC, the video monitor, the printer, the scanner, or other peripherals.

One other investment you can make as you go along is to put labels on your wires. Even a Wi-Fi network may have a dozen or more AC, power adapter, peripheral and LAN cords. Labels may seem like a waste of time at first, but the more wires are added to the mix, the more they resemble each other. The labels don't have to be pretty or perfect, merely legible. White electrician's tape marked with a permanent pen will suffice. The hours you spend will be recovered the first time you have to take your network apart.

Here's another general installation tip: Don't try to do it all in one night. Begin with the expectation that you'll spend an hour or two, and then pick it up the following day, as well as the next. Forcing yourself leads to fatigue, frustration, and sloppy work that comes back to haunt you. If you hit a snag along the way and it's late, there is no dishonor in setting it aside and hitting it again when you are fresh. Often a new approach will suggest itself to you while you are relaxing. The whole process will be more enjoyable.

The install kit should contain the other parts you will need:

- A step-down transformer to provide low voltage power to the modem
- A short LAN cable to connect the modem to your computer's Ethernet port
- A telephone cable to connect the modem to a phone jack
- Several microfilters, if you are using a Telco DSL modem
- Software, on CDs or floppies
- Possibly, an Ethernet adapter card, which you will have to install if you don't have one already
- Written instructions

New residential DSL users are sometimes provided with a combined DSL modem in the form factor of a PC add-on card. These are known as PCI DSL modems and are fine if you only want to connect one PC. Similarly, a modem USB cord only services one PC. Otherwise, your PC will have to be prepared in advance with a Network Interface Card. Most laptops have them built in. Some ISPs will include them with the other self-install hardware. Fortunately, these interface cards are relatively inexpensive these days, should you have to buy one. Prices range from $20 to $60, and you won't need fancy add-ons. If you are using a tabletop PC, you will have to open it up and install the card yourself.

Newer NICs move data as fast as 100 Mbps, which is considerably faster than the 10 Mbps (10BaseT) socket on your modem. Wi-Fi access point/routers have at least one wired outlet for connections to existing LANs, and these ports run at 100 Mbps.

We will assume, given this book's title, that you plan on installing a Wi-Fi network. It is still a good idea to set it up initially as your DSL provider expects. That is, as a single PC wired directly to your DSL modem. Starting this way, instead of cabling everything at once, enables you to build up your network in simple stages. You will have confidence that the next addition will work if you know that you are building on a working foundation. Going at it in a step-by-step fashion makes troubleshooting much easier, because you know that the defective piece of the puzzle is the one you added last. And finally, if something does go wrong with your home network, the DSL provider will only troubleshoot their portion of the network.

Your tech support advisor will insist that you disconnect any routers, access points, or firewalls so that the ISP's technician can concentrate on their own equipment (see Chapter 6, A Case History).

After you have installed the NIC card into an empty PCI slot on your PC and reassembled your computer, you will have to install the software given to you by the card provider, whether that is the manufacturer or your DSL company. This consists of inserting the CD-ROM sent with the card and following the instructions. You are going to have to know some unique identifying numbers:

- Your own IP address, if you have been given a static address. Set the card for dynamic addressing or *Dynamic Host Configuration Protocol* (DHCP) if not.
- Your subnet mask.
- The interface's gateway address.
- DNS addresses.
- Login type: PPPoE or *PPP over ATM* (PPPoA).
- Login name and password.

You may be able to get these from your ISP in advance of your service activation, so that you will be ready to go when the day arrives.

Telco technicians will install full-rate DSL connections, but the home version you will buy will involve a little labor on your part. The self-install kits shipped along with residential DSL modems do not require the use of tools. One of your tasks will be to plug a microfilter into every active telephone jack to isolate the DSL signal from your telephones. Empty phone sockets do not need microfilters, nor does the socket serving the DSL modem. If you have two or more telephone lines, you need only install filters on the line that will be serving the DSL modem. Microfilters resemble foot-long extension cords for telephones with a lump at the end (see Figure 2-5). Their purpose is to pass voice, but block high-frequency data signals, leaving those exclusively for the DSL modem to use. It's a good idea to count the phones, fax machines, and answering machines that will reside on the same line as your DSL modem and tell your DSL provider in advance how many microfilters you will need. Every item on the shared line will need a filter, including the satellite TV decoder's phone line. You can also buy more yourself from electronics stores if you are in a hurry or you add a telephone.

Microfilters also protect certain telephones. Although you cannot hear noise at 4 kHz, it has been known to produce static and whining in some telephones. Also, the data signal can confuse phones equipped with automatic level controls. Check the phone for the presence of a dial tone after

Figure 2-5
A microfilter has a male telephone (RJ-11) plug at one end and a socket at the other. Photo by Daly Road Graphics.

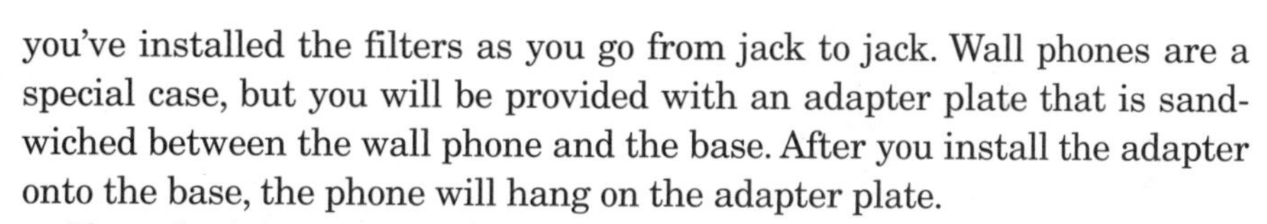

you've installed the filters as you go from jack to jack. Wall phones are a special case, but you will be provided with an adapter plate that is sandwiched between the wall phone and the base. After you install the adapter onto the base, the phone will hang on the adapter plate.

If you don't have spare phone jacks near your PC, you'll be provided with a two-for-one splitter. This small adapter simply converts one plug into two sockets. You use one of the sockets for the DSL modem without any filters and the other with a microfilter for the telephone. With everything set, installation is fairly simple:

1. Turn all the power off.
2. Plug one end of your short Ethernet jumper cable into the socket on your NIC card in your PC.
3. The other end goes into the corresponding socket on the modem. This can be labeled LAN, Ethernet, or 10BaseT. If you are using a model equipped with a USB cable, plug that into an open USB port in the PC.
4. Plug one end of the RJ-11 (phone) wire into the socket on the modem. This can be labeled line, WAN, or DSL. The other end goes into a phone jack (or the two-to-one adapter) just as though the modem were a telephone.

5. Plug the cord from the power adapter transformer into the modem (see Figure 2-6).
6. Plug the transformer into a live wall outlet. Turn on the modem.
7. The power indicator on the modem should light. Then other indicators will blink to show that the modem is testing itself and trying to connect to the telephone company equipment or the cable company. Check the instructions that came with it to see what the lights mean and what they should normally look like. If the sync light comes on, your line may be live.
8. Turn on your PC. After it boots, if your NIC is working properly, it should show a green indicator light, meaning that it "sees" the DSL modem. The modem should have a similar indicator to show that it sees the NIC card.

In addition to the software drivers and IP protocol stacks that you installed to make the NIC card operate, you may also have to install software specific to your ISP. Many of them use a version of the same protocol designed originally for telephone modems, known as PPP. Your DSL variant takes these packets and sends them over an Ethernet connection, thus giving us the protocol PPPoE. This is not included as part of the standard Windows package. If your ISP uses this method, you will have to install the software for it as well, independently of your NIC. The software package enables your DSL modem to connect when you first turn it on, since your PC will have to supply a login name and password, just as for dial-ups.

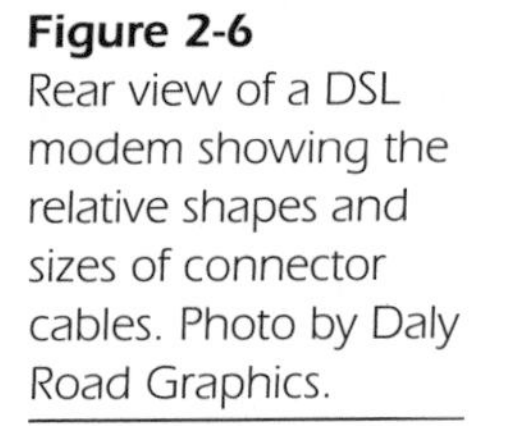

Figure 2-6
Rear view of a DSL modem showing the relative shapes and sizes of connector cables. Photo by Daly Road Graphics.

You may be tempted to try out your connection in advance of the due date if everything is in place. It may work, although you have no reason to expect it to. If you've been used to a telephone dial-up, you will be pleasantly shocked at how fast your new high-speed connection runs. It's not as quick as industrial links, but subjectively it will be hard for you to tell the difference between it and a 10 or 100 Mbps connection, because content providers usually can't deliver data much faster than you can receive it at home.

After your broadband PC is running, here are two cautions. First, don't throw out your old telephone modem. If you lose your DSL connection, the dial-up will be your only door to the world and a troubleshooting tool as well. For example, your dial-up is a simple way to determine that your login and password are still valid. Secondly, as you graduate to a Wi-Fi LAN, don't throw away your PPPoE software. It will peacefully coexist, unused, on your laptop or other PC. You will need that for troubleshooting as well. Your DSL and/or ISP will insist that all unknown or (as they view it) nonstandard equipment is taken out of the connection before they even start to investigate.

After comparison shopping you should be able to pick the best way to make your home one strand of the World Wide Web. Research is the first and most important step toward building your home Wi-Fi network. In the next chapter we consider various means of spreading the wealth of Internet data around the rooms of your home. That will require a little labor on your part also, but fortunately, it's not that tough to do.

CHAPTER 3

From the Doorstep to the Tabletop

After your broadband connection to the Internet is established, you may share it among several home-office PCs. This chapter details the benefits and methods of sharing data among PCs in a home network. We provide examples ranging from the simplest, such as peer-to-peer networking using infrared ports, to a complex Wi-Fi infrastructure. We will also help you choose a distribution medium.

Sharing a Broadband Internet Connection

If you have an existing network, and you've just installed a broadband link to the Internet, you can make that link work for all your existing computers, whether the link is via satellite, cable modem, or Telco *digital subscriber line* (DSL). You should check with your *Internet service provider* (ISP) before you do so, however, because their policies vary. Most don't mind customers sharing access among their own PCs. Most do mind customers sharing with friends or neighbors. Some don't care, and some flatly forbid it and actively seek out users who share in any way. As mentioned in the previous chapter, you should base your choice of ISP on this policy question as well as on other factors.

The simplest way to share an Internet line is to purchase additional *Internet Protocol* (IP) addresses from your ISP. No providers complain about this method, because they make a profit on the exchange. Prices may vary from $3 to $10 per month, assuming that your ISP is willing or able to sell extras. You will need one for each PC you plan to connect. Afterward, all the PCs will have to be interconnected through a hub or switch, which will also have a line to your broadband modem.

It is possible to achieve the same end by using a clever device that selectively channels traffic between the broadband link and your client PCs. It is called a *router*. This function can be carried out by a dedicated PC running special software or by a piece of special-purpose hardware. Many hardware routers are made more cost effective by including built-in firewalls and high-speed switches, which makes them an increasingly popular solution for broadband sharing. Software and hardware routers are *Network Address Translation* (NAT) devices. They emulate a single PC when they connect to your ISP. They will even perform an auto-logon if they have to. Afterward, any traffic destined for a PC in your home is directed (routed) to it as though it were the only PC, even though you may have several. Similarly, any traffic from any PC to the Internet gets funneled into one DSL line. This provides an extra measure of security for your PCs because to someone on the outside it appears as though only one IP address is active. In this way, a NAT router acts as a firewall.

Another measure of security is gained from the range of IP addresses distributed to the clients, because they belong to a special class:

- 10.0.0.0 to 10.255.255.255
- 172.16.0.0 to 172.31.255.255
- 192.168.0.0 to 192.168.255.255 (the most common)

Addresses in this range function like any others, with the exception that Internet routers won't pass them along. This means that your PCs are more isolated and therefore more protected. A hacker trying to access any of your addresses within these limits will have no luck, as his or her packets will hit an immediate dead end. DSL and cable are full-time Internet connections. They make your PCs easier to find and more vulnerable to attacks from malevolent hackers than they were on your old dial-up connection. This is especially true if you have a static IP address since a hacker can automatically "bookmark" your home network and return to it at leisure.

Software Routers

If you have more PCs than money, you might choose to use one of them as a *software router* or, as it is sometimes also called, a *proxy server*. If you use a PC in this way, it must stay on for as long as you want to use any of the others. It is generally slower than hardware routers. Remember that all your clients should have their own firewall and virus protection software installed. Usually, the host computer that runs routing software will have two network cards: one for the DSL/cable modem and the other for your *local area network* (LAN). You can buy several software packages for this job:

- Wingate from www.deerfield.com
- Sygate from www.sybergen.com

For Macs:

- IPNetRouter from www.sustworks.com
- SurfDoubler from www.Vicomsoft.com

Prices for the others can range from $35 to $700, depending on the number of users and other add-ons, such as parental controls. But you may already have a software router. Windows 98SE and Windows 2000/XP include *Internet Connection Sharing* (ICS) as a bundled function. If your PC does not already have it, you can't download it from the Microsoft web page, although those pages are a good source of advice and software updates.

Connection Sharing, Windows 98/ME ICS provides for sharing a single Internet connection to a small peer-to-peer network such as with Wi-Fi's *ad hoc mode*. ICS requires a host computer as a *Dynamic Host Configuration Protocol* (DHCP) Server that dynamically hands out IP addresses to as many as 10 other network PCs as they boot up and ask for them. It also includes the functions of a *domain name server* (DNS) proxy. (We cover ICS configuration in more detail in Chapter 4, "Connection Sharing for XP.")

When one of your clients refers to a web site by its alphabetic name, the Internet must first translate it to an IP address, as you might do with a telephone directory. These requests must be shuttled back and forth through the server PC along with other traffic. The ICS server also uses *application programming interfaces* (APIs) to aid configuration, report status, and manage the dial-up. (Yes, you can use it with a telephone modem if you must.)

Hardware Connection Sharing

One of the killer applications for home Wi-Fi is sharing connections to a broadband Internet link. To do this, you will have to set up your wireless workstations in *infrastructure mode*. This means that in order to communicate to each other, they will all first go through a wireless hub. This connects them to each other, and the built-in router will in turn connect them all to your cable or DSL modem.

Thanks to a hungry mass market and economies of scale, the cost of hardware routers has dropped, often to a price lower than software-based equivalents. These intelligent devices do everything that the software routers can do. What's more, modern hardware routers usually have other functions built in that otherwise would require you to make several purchases and to integrate several pieces of hardware afterwards. Most Wi-Fi routers usually include four or more RJ-45 sockets to accept and interconnect traffic from your wired PCs or your existing LAN hubs (see Figure 3-1).

Figure 3-1 Rear view of a combination access point/router/firewall/ LAN hub. Inputs from local and distant sources are intelligently channeled for maximum through-put and security. Photo by Daly Road Graphics.

These ports are *switched*, meaning that traffic is selectively channeled between them so that any one port sees only the data destined for it. The older hubs acted like a party line, with heavy traffic on one echoed onto all, thus sapping their capacity. The built-in switched ports auto-sense speeds of 10 or 100 Mbps and interface between computers or networks running at different speeds. Some routers include ports for printers, enabling you to share one printer among several home-office PCs, wired or otherwise. Some have built-in telephone modems that will dial and share a connection to your ISP when the DSL line fails.

Home routers come in different flavors, some having specialized built-in interfaces that connect directly to existing "no new wires" media, such as phone- or power-line home networks. It is possible to combine interfaces, thereby adding a Wi-Fi cell to a phone-line network that includes a shared Internet link. For this task, you will need an *access point*, which is a simplified version of a Wi-Fi router. Access points have only one wired port. Their purpose is to provide a pathway from a Wi-Fi cell, making it appear as a wired PC to a preexisting wired network.

Alternatives to Wi-Fi

In some circumstances, you might not be able use Wi-Fi in your home. You may have PCs that are so spread out that one access point will not cover them all. You may have a remote building, such as a barn, garage, or shop that you'd like to include. You may have already installed an isolated wired home network of some kind. The term *wireless* usually refers to radio, but it can also be used to mean "no new wires." Other ways exist for connecting your PCs without resorting to planting new utility poles on your property or drilling holes in the walls. You can interface these other media to a Wi-Fi access point to utilize the advantages of all of them. We'll start with the simplest.

Infrared Links

Many PCs and *personal data assistants* (PDAs) are equipped with infrared ports, which transmit data between themselves in the same fashion that your TV remote control uses: over a modulated beam of invisible light. These are fine for communicating from one to another, as long as they are in a line of sight, usually sitting right next to each other. Bright sunlight, however, can interfere with the signal.

You can buy infrared adapters based on the *Universal Serial Bus* (USB) for desktop PCs, and some adapters will pick up diffused signals that have bounced off a wall. But infrared ports cannot be used around corners or from room to room, as is the case with Wi-Fi. This also means that someone outside the house cannot pick up the signals, as is the case with Wi-Fi. The transmission speed varies between 1 and 4 Mbps, depending on the manufacturer and on computer placement. The process of transferring files over a light link is known as *beaming,* and updating files back and forth is called *synchronization*.

Telephone-Line Home Networks

If you are reluctant to install computer cables, you can utilize the ones that the telephone company has already put in to connect your PCs. Your existing phone lines can be made to carry data as well, using the same technique (but not the specific audio frequencies) that DSL employs. The process of using an impressed carrier signal to interface PCs in this way is known as the *Home Phoneline Networking Alliance* (HomePNA) method. With a HomePNA network, you can talk on your telephones and share resources over a network at the same time.

The HomePNA interface cards will not interfere with a DSL connection. About 50 devices can be connected to a phone-line network within a home while still maintaining the 10 Mbps speed. Additional devices can also be added, but may result in overall slower network speeds. The maximum separation between computers is about 1,000 feet. Your network should cover less than 10,000 square feet, an area that is larger than most homes.

How HomePNA Works Home phone-line networks use an inaudible carrier signal on the line that (usually) doesn't interfere with voice communications or faxes. Specifically, the technique is called *Frequency Division Multiplexing* (FDM). FDM puts computer data on separate, discrete frequencies intended to avoid interference. The protocol involved is a derivative of Ethernet, using a proprietary compression technique. It also uses Ethernet's method of avoiding interference from multiple stations sending at once. This is abbreviated with the lengthy acronym CSMA/CD, which stands for *Collision Sense Multiple Access with Collision Detect*. Simply put, each station listens for a clear channel before it starts transmitting. If the line is busy, the station will wait a random interval before trying again.

The technical challenges to overcome are considerable, given the medium involved. Unlike CAT5 wired networks that are built to exacting standards, a phone-line network can change randomly. Users can plug and unplug telephones and other devices, complete with extension cords. Every time this happens, another "branch" is grafted onto the phone-line wiring "tree." A transmitted signal is dampened and scattered as it reverberates on the wiring. In addition to added loads, the lines are sometimes unbalanced, or *unterminated*, as some phone sockets have nothing plugged in. The lines pick up high levels of noise from appliances, heaters, and air conditioners. Finally, the interfaces must peacefully coexist with the customer's existing equipment and with the telephone company's gear, as mandated by the *Federal Communications Commission* (FCC). Nonetheless, it works well in 99 percent of the homes in which it is employed.

Network adapters for your *home area network* (HAN) are available in the usual variety of form factors: internal bus *network interface cards* (NICs) and USB connectors, but also the less common parallel port connectors. They are priced about the same as equivalent Ethernet components. Every computer so networked must be placed near a telephone jack. If you aren't totally opposed to running new wires, you have the option of running telephone cable from one room to another through the walls or along baseboards. You also have the option of using a telephone extension cord to daisy chain your PCs together and not use the in-wall wiring.

One of the attractions of this method is its simplicity. You just connect a regular (RJ-11) telephone cable between the NIC or USB adapter to the telephone wall jack (see Figure 3-2). You can continue to use your wall jack for a telephone with a *modular duplex jack,* sometimes called a Y-connector. Another advantage is that it is relatively inexpensive. If all you want to do is interconnect a pair of PCs, you can do so for $50. Gateway devices are available to connect a telephone-line network to a cable or DSL modem.

Many home PC users protect their equipment, especially telephone modems, from voltage spikes by using a telephone line surge suppressor (see Chapter 5, "Power Corrupts"). Unfortunately, these devices will suppress the 2 Mhz HomePNA data signal also. As mentioned, telephone lines carry noise at this frequency from multiple sources, which can cause the data stream to bog down. If the problem is unusually intense, you may need to buy special *low-pass filters*. These are similar in shape, installation, and function to the DSL microfilters described in the previous chapter (see Figure 2-5). They permit voice but stop noise from cordless phones or fax machines from entering your HomePNA network. If your home uses a digital PBX system, it will not be compatible and you will have to choose another networking method.

Before you buy, make sure the equipment conforms to the HomePNA standards (www.homepna.org). The HomePNA is a trade association composed of over 90 major telephone-wiring technology manufacturers. It seeks to put a wider array of peripherals on telephone lines, such as printers, streaming voice and video, and home automation devices. As with the Wi-Fi Alliance, HomePNA develops common standards to promote interoper-

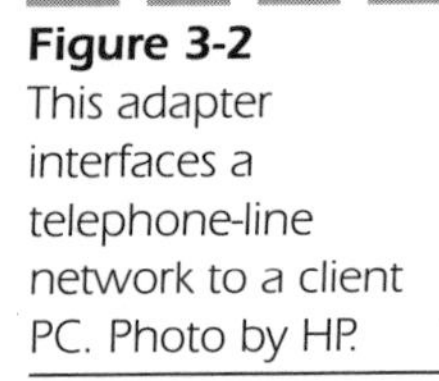

Figure 3-2 This adapter interfaces a telephone-line network to a client PC. Photo by HP.

ability between different manufacturers. The equipment is available in 19 European nations as well as North America.

The first of their standards was version 1.0, which transmitted data at a rather slow 1 Mbps. The latest is 2.0, originally developed by Broadcom, which sends at 10 Mbps. The faster adapters will work with the slower ones, which are still on the market. A newer *third-generation* (3G) HomePNA specification promises to deliver *quality of service* (QoS) at a throughput rate of up to 100 Mbps. HomePNA equipment has the following characteristics:

- It is standardized and interoperable, reliable, easy to install, and relatively cheap.
- If you don't mind peer-to-peer mode, you can use it without buying other gadgets, such as hubs.
- You can interface it with other network technologies.
- It is available for Macs and older Wintel PCs.

Networking over Electrical Wiring

As with DSL and phone-line networking, power-line technology uses your existing AC wiring as a transport for a data carrier frequency. Power-line networks use an exclusive set of radio frequencies that won't interfere with remote-controlled on-off switches. The raw data rate is about 20 Mbps, but error correction and other overhead subtracts from that, leaving an actual rate of about 14 Mbps. It won't affect your electric bill and it's even more convenient than using phone lines, because you probably have empty electrical outlets all over your house.

The hardware typically consists of an adapter that attaches to the computer, usually through the parallel port or a USB connector, and a proprietary interface that plugs into the AC outlet. The adapter will have its own surge protection built in. You will need one adapter/interface set for each PC you intend to connect.

As with Wi-Fi, privacy is a concern with this technology. The signal can migrate through the incoming power lines to other nearby homes. It will not jump an electrical distribution transformer to the world at large, but as many as six other homes may be tied into a single distribution leg, and more than that may exist in an apartment complex. The power-line standard, however, does include packet encryption.

How AC Line Networking Is Done Power lines are even noisier than the telephone lines described and are extremely noisy under the best of circumstances. They change constantly too, as customers plug in some appliances and remove others. Some of the devices being powered are inherently noisy by themselves, such as fluorescent lights, switching power supplies, and dimmer switches. The circuit breakers in power panels are signal sponges. HomePlug technology deals with the hostile data environment by adapting to changing conditions, pushing high throughput on some channels, and slowing down on others to plow through noise. Also, it uses thorough error detection and *automatic repeat requests* (ARQs) to ensure that the line appears reliable to the driving software. The HomePlug specification uses the same *Orthogonal Frequency Division Multiplexing* (OFDM) that is used in the newer 802.11a Wi-Fi networking standard. Basically, the HomePlug specification works by sending most of the data on the clearest of 84 channels (between 4.3 to 20.9 MHz), dynamically shifting data to alternates if some of them become swamped with noise. The data signal has no effect on your home's electricity, which is immensely more powerful. The age of the wiring in a house does not appear to be a factor.

The software will automatically detect your plugged-in nodes. Adapters are available to provide printer sharing, and routers are available for interfacing multiple home PCs to a single Internet link. Many adapters have AC sockets built in, so that you won't lose a place to plug in a desk lamp when you plug in an adapter. Depending on the brand, the resulting networks can be client/server or peer to peer. The advantages for using your power lines for networking are as follows:

- You already have multiple AC outlets in every room.
- It won't interfere with other home networking technologies.
- It's cheap.
- It works the world over, even on older wiring.
- It's easy to install.
- It's easy to add more nodes.

Before you buy it, make certain that it has been tested as conforming to the specifications of the HomePlug Powerline Alliance (www.homeplug.org). As with the Wi-Fi Alliance and the HomePNA organizations, the Powerline Alliance conducts tests to guarantee interoperability between manufacturers.

True Wireless

Some of the following material was summarized in Chapter 1, "Is This Trip Necessary?," but since Wi-Fi home networking is the subject of this book, we will go into more detail here.

Wireless LAN (WLAN) products have become widely popular and firmly established in the marketplace. In late 2002, Microsoft announced their line of home wireless equipment (see Figure 3-3), thus confirming its popularity and permanence. Wireless networking has been around for a decade, but has only achieved wide popularity in the past few years. This is for two reasons. First, the early attempts, such as the HomeRF standard, had slow transmission speeds and used a variety of proprietary protocols. You could not mix one manufacturer's hardware with another or be assured that your investment would be useful in the years to follow. The most common wireless solution for home and business networking today runs at a speed of 11 Mbps and does adhere to international specifications. If you buy an access point from one manufacturer, it will work with the built-in Wi-Fi interface in your laptop, assuming that both have been constructed to the standards agreed to by the *Institute of Electrical and Electronics Engineers* (IEEE).

The IEEE Standards The IEEE is the world's largest professional organization, with over 320,000 members worldwide. They sponsor technical

Figure 3-3 Microsoft introduced a Wi-Fi-compatible product line in 2002. Photo courtesy of Microsoft.

conferences and meetings and publish white papers. They also take input from many professionals, scholars, and students and resolve those into *standards*, which are detailed rules as to how a particular piece of electronic gear should work. Perhaps their most significant standards for our present purposes involve specifications for LANs. These rules all start off with 802.*XXX*, as does the particular subset dealing with WLANs.

Equipment makers have found through trial and error that they have a better chance of selling to a large market if their equipment can be easily judged against competitors and even be successfully intermixed. Those who go it alone are sometimes said to be engaging in a *connector conspiracy*, which means that they hope to grab and hold you as a customer by selling you something that will only work with other equipment that they sell. Some manufacturers genuinely believe that their new product is completely new and superior. Some of them actually manage to have their device accepted as a new standard, but they are taking an awful chance when they do. The slang term "to be Betamaxed" means to have a superior technical method eclipsed by a competing technology that is either superior for other reasons or simply better sold. The classic example is Sony's introduction of the Betamax videotape format. Years later, owners of the Betamax machines found themselves left high and dry as the world settled on the competing VHS format. (Sony stubbornly continued to manufacture the home version of the machine until mid-2002.)

The IEEE exists partly to keep that from happening by proactively taking the best aspects of competing technologies and combining them into one standard that everyone can live with. Other than market forces, they have no legal means to enforce a standard once they have handed it down. In fact, some wireless home networking manufacturers still exist whose equipment is close to the standard and works well in most respects, but for some reason will not work interchangeably with all of it.

Whichever Standard You Choose Once a standard has been set, it does not change, but real-world circumstances change all the time. Technologies, especially new and popular technologies like Wi-Fi networking, are constantly being improved and applied in new ways to solve new problems. As the number of Wi-Fi technology users jumps upward from the 20 million on the air in 2002, for example, interference between users is bound to become a common problem. This is driving a migration to less crowded frequency bands. The IEEE must move pretty quickly to accommodate changing applications and a growing market by adapting existing standards and, if necessary, making new extensions to them. Their goal is to allow for new technical capabilities without rendering an existing class of equipment obsolete. The result of their efforts is a continuous stream of specifications for new variations of equipment, all beginning with the number 802.

As the list of rules gets longer, it inevitably becomes more complex and more confusing. Even those who deal with it daily as a condition of employment refer to it as an alphabet soup. Some manufacturers add features that have not been tested or approved to gain an advantage. But in order to remain in compliance, their equipment must be smart enough to communicate with any existing equipment that does adhere to the published standard. Examples are wireless interface cards that will move data twice as quickly as nonwireless cards at 20 Mbps. When they encounter an interface broadcasting at the 802.11b standard rate of 11 Mbps, they must slow down to match it and do so without user intervention. One manufacturer's turbo mode might not work with others, even if both will interoperate at a standard speed. Eventually, the IEEE might define a technical standard that allows for turbo, and one or both of the manufacturers will have to give way in order to comply.

All these standards utilize very high frequencies on unlicensed portions of the radio spectrum, designated by the FCC as the *Industrial, Scientific, and Medical* (ISM) bands. They all use relatively low power and yield a relatively short range when compared to a cell phone. All use a frequency-hopping method that minimizes interference and provides a basic level of privacy. We will try to explain the entries as they existed in mid-2002.

802.11: The Original This standard was approved in July 1997 and governed WLAN operations at the relatively slow data rate of up to 2 Mbps at 2.4 GHz.

802.11a: A 5 GHz Fast Lane 802.11a is one answer to overcrowding in the lower-frequency bands. The other standards, which operate at 2.4 GHz, must share the road with wireless telephones, baby monitors, and microwave ovens. In the 5 GHz *Unlicensed National Information Infrastructure* (UNII) spectrum, however, equipment suffers less interference, partly because it uses a specialized encoding method (OFDM) to make it so. The higher frequency also enables a higher set of data rates, from 6 to 54 Mbps depending on the signal strength. These two advantages make the standard a good choice for supporting streaming audio and video. Some futurists envision our homes equipped with one centralized media box getting input from the Internet and wirelessly driving an array of portable audiovisual display devices. The range for this type of equipment is 50 to 200 meters.

The standard was approved in September 1999 and products began shipping in early 2002. It improved on the original 802.11 standard, which moved data at a slow one or two Mbps, using a modulation technique called *Complimentary Code Keying* (CCK). The modern version provides a top rate of 11 Mbps in the 2.4 GHz frequency band. Some manufacturers are producing equipment that will take input from either band and interface the two.

802.11h: A Noninterference Directive The 'A' standard described previously can only currently be used in North America and Japan because it interferes with satellite communications in Europe. European satellites are designated as the primary use of the band. The h standard compensates in two ways: by dynamically adjusting transmit power to the minimum level necessary to get through and by dynamically selecting frequencies to avoid stepping on one that is already in use. Fortunately, it should be backward compatible with existing a-type equipment. As h sales increase in Europe, the cost per unit should drop worldwide. This standard is still under construction by an IEEE *Task Group* (TG).

802.11b: Your Safest Bet Today the most popular standard by far is known as 802.11b, which was finalized in September 1999. It improved on the original 802.11 standard, which moved data at a slow 1 or 2 Mbps, using a modulation technique called *Complimentary Code Keying* (CCK). The modern version provides a top rate of 11 Mbps in the 2.4 GHz frequency band. Like the others, if the signal degrades, the equipment will shift to lower speeds to maintain a connection in a process known as fallback, rollback, or scaleback. It is more prone to interference than 802.11a, but the lower frequency gives it a longer range estimated between 75 to 300 meters. Equipment of this type is most likely to be Wi-Fi certified and have the largest number of ready-to-go subscribers. Recently, laptop and handheld manufacturers have started to include built-in radio interfaces of this type as standard equipment.

802.11c: Wireless Bridging 802.11c deals with *bridging* and applies only to wireless access points and bridges. It is not relevant for home Wi-Fi users.

802.11d: One Big Happy Family 802.11d and h accommodate European and Asian governmental regulations governing radio devices. The use of frequencies, particularly in the 5 GHz bands, varies widely from one country to the next.

802.11e: Quality of Service (QoS) Existing Wi-Fi standards don't handle multimedia well. Without priority for streamed packets, annoying pauses in voice or video can occur while a WLAN moves email or other data files first. The IEEE's e group is trying to determine a better version of the packets' *Media Access Control* (MAC) layer to improve the QoS of streamed multimedia applications, such as MPEG-2 video. They also hope to improve authentication and security.

When the standard is finalized, you may be able to upgrade your existing equipment with firmware patches instead of having to replace it. Newer models using this standard should be backward compatible with older ones that lack it.

802.11f: Interaccess Point Protocol The chief advantage to Wi-Fi LANs is the mobility they grant users. But as users move from one room to another, they may need to have their data streams rerouted through a different access point. To coordinate the handoff, the two fixed access points have to communicate to each other. This means that access points from different vendors may not interoperate when supporting roaming. Most of them do, especially if they've been Wi-Fi certified, but 802.11f specifies an interaccess point protocol that guarantees it. The standard is under construction as of October 2002.

802.11g: Officially Sanctioned Turbo Mode The 802.11g standard will be released for the 2.5 GHz band in 2003. Running at 54 Mpbs, the G equipment will be faster than the current B models and less susceptible to radio noise because it will use the newer OFDM transmission method. Its greater capacity makes it a contender for the job of wireless multimedia streaming.

In order to make the g standard backward compatible though, some technical issues will have to be resolved. Since the two types cannot intercommunicate directly, the b equipment must rely on the transmission of *request-to-send/clear-to-send* (RTS/CTS) packets to avoid covering a 802.11g signal. This generates overhead and lowers throughput for users of both types. The draft version of this standard was published in late 2001, with final ratification expected in late 2002 or early 2003.

802.11i: Security Fixes One chief drawback to Wi-Fi LANs is the scheme originally devised to provide security for users. This is a serious concern because transmissions wander through walls and floors. The IEEE is devising enhancements to bolster *Wired Equivalent Privacy* (WEP; see Chapter 4). This data encoding method uses static encryption keys, making it possible for knowledgeable hackers to decode intercepted data. 802.11i will use more complex encryption techniques, such as the *Advanced Encryption Standard* (AES). You may have to download new drivers or firmware from your manufacturer's web site, or you may even have to replace your equipment with newer models in order to take advantage of this new standard when the TG has completed work on it.

802.11: Next Generation The *Wireless Next Generation* (WNG) specification seeks somehow to combine and streamline all of the above into one universal standard. In addition to the previous TGs, the 802.11 work group is studying new methods to increase throughput and more efficiently use the available radio spectrum. One example of this is *Ultrawideband* (UWB) modulation.

Ultrawideband (UWB) Unlike the 802.11 standards described, which operate within the confines of a narrow spectrum, UWB transmits a series of very narrow, low-power pulses. UWB was granted a limited license by the FCC in February 2002 for use in the 3.1 and 10.6 GHz bands, but only indoors or in handheld peer-to-peer applications for now. It can be used to find underground pipes or buried coal miners, but UWB can also be used to send data. Thus, it seems like the ideal channel for wireless TV, HandyCams, or DVDs, which require up to 12 Mbps to get their images across. The new *high-definition TVs* (HDTVs) need almost twice that much bandwidth.

Bluetooth *Bluetooth* (BT) is a communications technology named after a tenth-century Viking king who had a sweet tooth for blueberries. BT uses the same ISM 2.5 GHz radio band as Wi-Fi and a similar frequency-hopping scheme to dodge fixed-frequency interference. BT has a 10-meter range and is primarily designed to provide what is termed a *personal area network* (PAN). It is one way to get rid of connecting wires to handheld or wearable devices, such as a wireless headset for a cell phone. Unlike a structured WLAN, these personal devices link up on an ad hoc basis. Some devices using the BT standard can interface a PC to a printer or to a wired network, just like Wi-Fi. Some equipment on the market today combines a Wi-Fi interface with one for BT.

Even so, BT and Wi-Fi are not compatible and will not communicate with each other. In fact, if you use both at once, older devices are likely to interfere with each other, with BT being the winner. Thankfully, technical adjustments approved by the FCC will alleviate this competition, and an IEEE TG is working on a standard (802.15 TG2) that will enable the two to peacefully coexist.

Some speculation in the wireless broadband industry still exists about which of these two technologies will dominate. It is most likely that they will differentiate instead, with Wi-Fi being the preferred method for connecting PCs to a LAN.

Don't Get Betamaxed Manufacturers are already designing equipment that will automatically shift from a to b to g, or even interface them all inside one box. Some newer laptops combine Wi-Fi and BT interfaces in this way. Customers can avoid being Betamaxed by assuring themselves that the LAN equipment they are going to buy has been tested and certified to work with other Wi-Fi equipment. An agency exists for just this purpose: the Wi-Fi Alliance. About half of the wireless equipment manufacturers have submitted their equipment to their labs for compatibility testing. An up-to-date list is maintained at www.wi-fi.org.

To recap, here are some reasons to go with Wi-Fi for your home network:

- You don't have to drill holes or pull wires through walls to install it.
- You can have a PC anywhere—in the backyard, if you like.
- It's easy to interconnect with existing Ethernet networks.
- It's easy to add users or widen the coverage area.
- It's efficient and constantly being improved. Newer versions will run faster.
- It is reliable.
- The range is between 75 and 300 yards depending on conditions.
- Wi-Fi hardware is interchangeable with other Wi-Fi hardware and versions.

- Wi-Fi works on many different kinds of PCs and operating systems.
- It's getting cheaper all the time.
- New hardware, particularly laptops, has Wi-Fi built in.
- You can get up, move around, and stay connected.
- You will be able to use your Wi-Fi laptop at the coffee shop, the airport, at school, and the mall.

Physical Planning for Your Wi-Fi Home Network

We will assume from here on that you've decided on Wi-Fi as the way to make your network. Even if you have an existing network using one of the other technologies, you can add Wi-Fi to it easily using CAT5 patch cords. RJ-45-type connectors are a common denominator for all home network types. A Wi-Fi router, for example, will typically have four RJ-45 ports to accept input from wired PCs or other wired LAN hubs.

As with the other wired methods previously described, every PC on your Wi-Fi network must have a specialized hardware interface. These come in different shapes and sizes to fit the different PCs they were designed to work with. Special-purpose printer-sharing interfaces are also available (see Chapter 1). Modern PCs, especially laptops, come pre-equipped for Wi-Fi from the factory. The interfaces perform the same functions as for their wired equivalents, but they also include a small high-frequency radio transceiver and antenna.

Ad Hoc Mode Wi-Fi can be useful even if your home remains cut off from the World Wide Web. You may want to wirelessly connect your local PCs, laptops, and printers in the simplest way, which is called *ad hoc mode* (see Figure 3-4). Each node communicates directly to every other node, as they might with infrared. But because the medium is radio, they can all send and receive from room to room or around corners. When sharing data in this fashion, your PCs become a peer-to-peer network. This is one of the advantages of Wi-Fi. A group of business travelers who meet in a lobby or a customer who walks into your home office can immediately become a part of a network you create at a moment's notice.

Figure 3-4 In ad hoc mode, every station connects to every other. Courtesy of Linksys.

The alternative method of organization is known as *infrastructure mode*. If your Wi-Fi interfaces are configured this way, they will look for a central point at boot time and use that to relay data to the other devices on the wireless network (see Figure 3-5). They will not communicate with the other stations, wired or otherwise, except through the wireless base. While doing so, they divide the base station's maximum throughput of 11 Mbps between them. Base stations can handle 50 or more wireless clients, but not if they are all trying to talk at the same time. Once the wireless router station receives data, it selectively channels it out of the wired ports at whatever rate the destination station can accept. If the wireless router station is sending to wired PCs, this is most likely 100 Mbps. If it is sending to the Internet over a DSL modem, the rate is 10 Mbps even though the modem can only communicate to the Internet at 1.5 Mbps. In every case, the chain of interfaces is only as fast as its slowest link.

Product Placement You should give some thought to your system's layout before you place the equipment. All your stations should be within the pickup pattern, but the pattern should not extend past the borders of your house. If you are starting from scratch, you can put your base station, be it access point or router, in a midpoint location. If you already have a wired network, then you will probably want to centralize your Wi-Fi base station close to the existing hub. Likewise, it is often easier to move a wireless base closer to a DSL modem than to move the modem. In any case, you can connect nonmobile devices to a distant base station with a 100 Mbps LAN cable (see Figure 3-6). A connecting patch cord's maximum length is 100 yards. That should be enough for all but the largest homes. Usually, long patch cords are unnecessary.

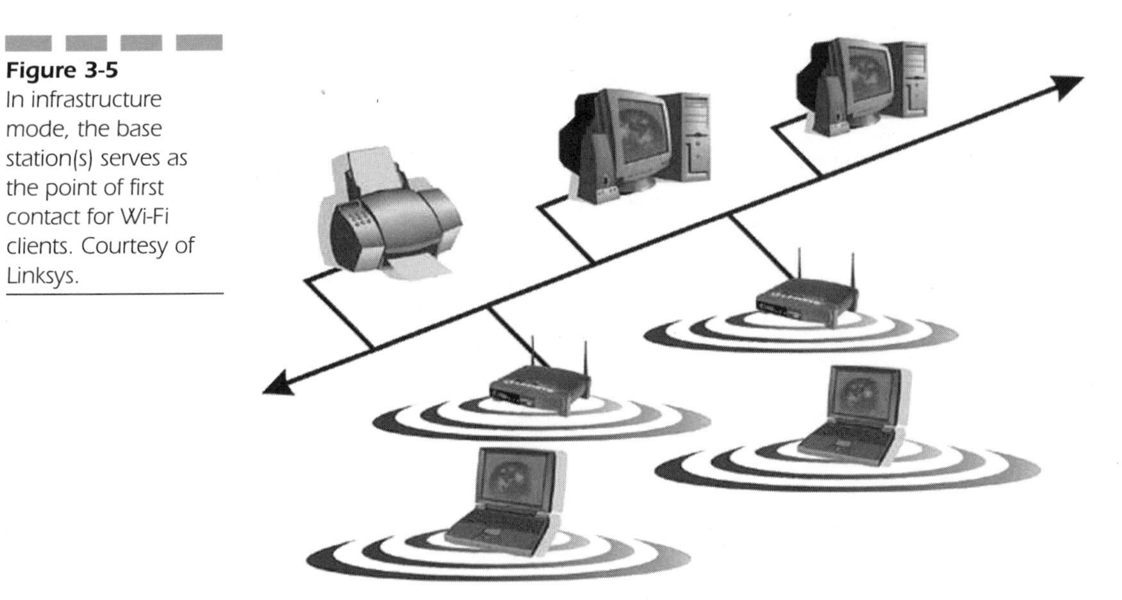

Figure 3-5
In infrastructure mode, the base station(s) serves as the point of first contact for Wi-Fi clients. Courtesy of Linksys.

Figure 3-6
Rear view of a Wi-Fi access point with the connections to wired PCs in place. The connection to the WAN (Internet) is on the far left. A dedicated back-to-back (rolled) port for an uplink to wired hubs is on the right. Photo by Daly Road Graphics.

Signal Strength and Range Remember that walls, ceilings, floors, and other barriers will interfere with the signal and decrease the range. Wi-Fi is sometimes called a two-wall technology for this reason. A typical line-of-sight range is given as 300 yards maximum, but it is actually hard to guess because the high-frequency signal bounces more than it is absorbed. At the very high frequency of 2.4 GHz,

- Vinyl walls are radio transparent.
- Drywall or wood paneling passes the signal, giving a range through five or six walls.
- Cinderblock walls allow enough signal strength for three or four walls.
- Precast concrete walls limit the range to one or two walls.
- A metal wall, metallic wallpaper, aluminum-covered insulation, or chain-link fences block the signal.
- Remember that most walls have metal pipes and wires in them. Stucco has embedded wire mesh.

Other factors can spoil your plans, but you can do some quick tests first to ensure that your home PCs will work on the first try. Laptop PCs often come with Wi-Fi hardware and software built in, so if you have a laptop or can borrow one, they make ideal, simple test instruments for your less mobile desktop PCs. Just call up the signal strength screen from the interface's configuration menu and check out the site you plan to use. The antennas in laptops are typically less efficient than those attached on desktop wireless cards. If you get a strong signal from a laptop at a given location, then you should have an adequate margin for your desktop station as well. Be aware that signal propagation on the 2.4 Ghz band is somewhat unpredictable, particularly at the weak fringes of the coverage pattern. As the signal degrades, the link will slow down to maintain the wireless circuit. Workstations that require high, dependable throughput, such as servers, should be closer together. Those that are used for slower applications, such as email, need not be so close.

Because the operating frequency is so high, it reflects off metal surfaces like a radar beam. Remember that the rear of most desktop PC chassis is metal (see Figure 3-7) so you may have to do some readjustment, particularly if your PC is backed up against yet another metal surface, such as a refrigerator or file cabinet.

Figure 3-7
A workstation's Wi-Fi antenna is parallel to a metallic surface. Photo by Daly Road Graphics.

One workaround for this problem is to use USB-compatible Wi-Fi interfaces for your desktops instead. These offer a few advantages over wired-in-place interface cards:

- They can be unplugged from one PC and relocated to another USB PC on the fly.
- The base antenna is at the end of a cord and can be easily moved for best coverage (see Figure 3-8).
- You can use it to interface a peer-to-peer Wi-Fi home network to a software Internet proxy.
- It's compact. You can use it in the corporate office and carry it home as well.

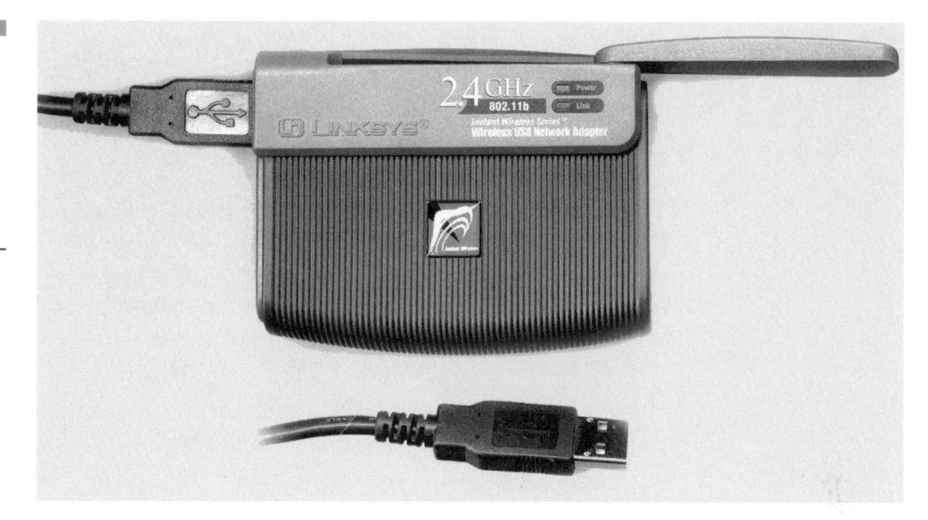

Figure 3-8 A moveable USB-based Wi-Fi network adapter with antenna extended. Photo by Daly Road Graphics.

Home Antennas

The twin antennas on access points and *Peripheral Component Interface* (PCI)-based interface cards can be tilted and rotated. Usually, they are most efficient when pointed straight up. If your laptop's cardbus interface is not getting a strong signal, you can often improve it by rotating the laptop or simply sliding it over a few inches. In theory, the signal will be strongest if the antenna points toward the access point or wireless router like a compass needle. Some of these cards now have more than one built-in antenna to avoid dead spots, but the rest have their built-in antennas oriented in the worst possible way: They are horizontal, and they are separated from the chassis of the laptop by half an inch or less. This sends the radiation pattern up and down, which is fine if you happen to be upstairs from your access point. If you are downstairs from one, you'll have a problem because access points have dead spots directly underneath. Apple's built-in interfaces deal with small-card coverage problems by placing a built-in antenna on the side of the *liquid crystal display* (LCD), which is vertical when being viewed. Other manufacturers are following suit.

If your environment is unusually difficult for some reason, you may have to resort to an external or add-on antenna, but lower-priced peripheral cards won't come with accommodating sockets or plugs. If you anticipate this problem, buy a NIC that does have an antenna socket. (In late 2002, Linksys introduced an add-on antenna range extender.) As standardized as these interfaces have become, none of the antenna jacks are of a consistent factor and intentionally so, thanks to FCC regulation section 15.203. It was intended to discourage experimentation by users. You must either buy the manufacturer's own brand of antenna or search for an adapter at an electronics store. (Table 3-1 lists some popular manufacturers and their types of add-on antenna connectors.) Unfortunately, the *radio frequency* (RF) signal is attenuated somewhat every time it must traverse an adapter.

It is natural for some folks to want to experiment and get the maximum range from their equipment. This is one of the primary motivating challenges behind amateur radio, for instance, but good reasons exist for getting a ham license and experimenting that way instead of doing so with your Wi-Fi equipment. The reasons against experimenting with Wi-Fi apparatus are as follows:

- Wi-Fi interfaces were not designed to be tampered with.
- Doing so violates FCC regulations.
- You might ruin the equipment if you don't make it less efficient.
- Letting your signal out lets others in.
- You don't need to. It almost always works fine as is.

Putting an external antenna on your roof to cure a few dead spots below is counterproductive. The signal from the interface will be soaked up by a long transmission line and be cut every time it must jump a connector or adapter. Omnidirectional antennas are least sensitive below. An antenna on the roof is subject to weathering, lightning, and water in the form of rain, fog, and condensation.

Table 3-1

Some Wi-Fi equipment manufacturers and the types of antenna connectors they've chosen for their equipment

Interface	Connector Type
Lucent/Agere/Orinoco/WaveLAN	Proprietary
Proxim RangeLAN-DS	RP-MMCX
Cisco 340/350 card, Samsung MagicLAN	MMCX
Cisco 350 AP, U.S. Robotics 2450	RP-TNC

Water as a Signal Sponge Water can be particularly troubling at 2.4 GHz. That particular frequency was originally chosen for microwave ovens because it was well suited for heating water, meaning that water absorbs it. Bolting a Wi-Fi antenna to your chimney may give you coverage all over the block on a sunny day, but hardly anything in a dense fog. Also, if it works well in winter, it may gradually go deaf in the spring, as the trees put on foliage.

Putting your access point next to a water cooler or an aquarium is an obvious mistake. Your fish will stay healthy, but a large tank will act like a black hole. Window plants and potted palms are also mostly water. So are people, which is why you can block the link from a laptop simply by resting your arm against the interface card.

Privacy Versus the Wireless Range Even if all your workstations have an adequate signal from your access point, one other check you should make with your signal-strength laptop is to walk around the periphery of your home to determine where the coverage pattern ends (see Figure 3-9). Believe it or not, too much signal spilling out to the street is not necessarily a good thing. It can interfere with neighbors' Wi-Fi networks, especially in

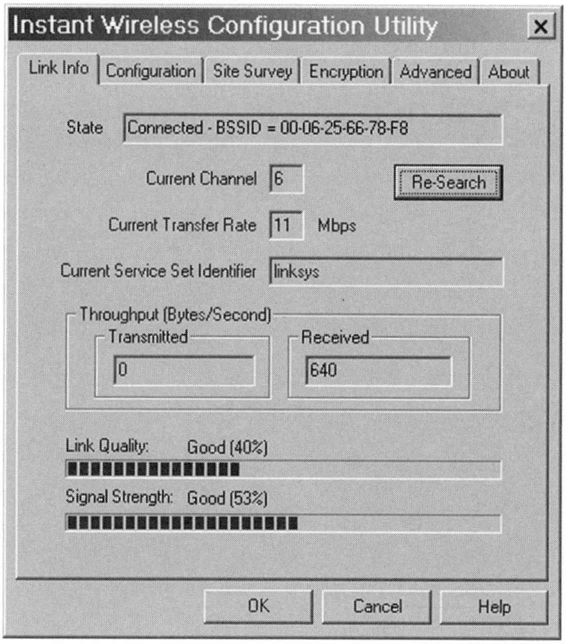

Figure 3-9
A laptop will tell you if your Wi-Fi signal is weak, strong, or too strong. Photo by Daly Road Graphics.

high-concentration areas such as apartment complexes. A full rate of 11 Mbps can be maintained even if the reported signal strength is less than half the maximum.

As mentioned repeatedly, a sloppy signal makes your access point more vulnerable to those on the outside. Most laptops come with built-in Wi-Fi interfaces these days, and many employees carry them home as tools for the job. So an increasing number of users are equipped to find your access point and use it if you allow them to. If you want to use a laptop at a great distance outside, temporarily place the access point antenna so that it looks out a window. It's more likely that you will wind up placing your base station so that its signal is deliberately *weaker*.

Wireless Bridges If you need to connect a distant office, a backyard workshop, or a separate wired network, you can use a wireless alternative to stringing a long wire between buildings. A Wi-Fi bridge consists of two specialized base units that work together as a radio "extension cord" for a wired network (see Figure 3-10). Their specific purpose is to extend the

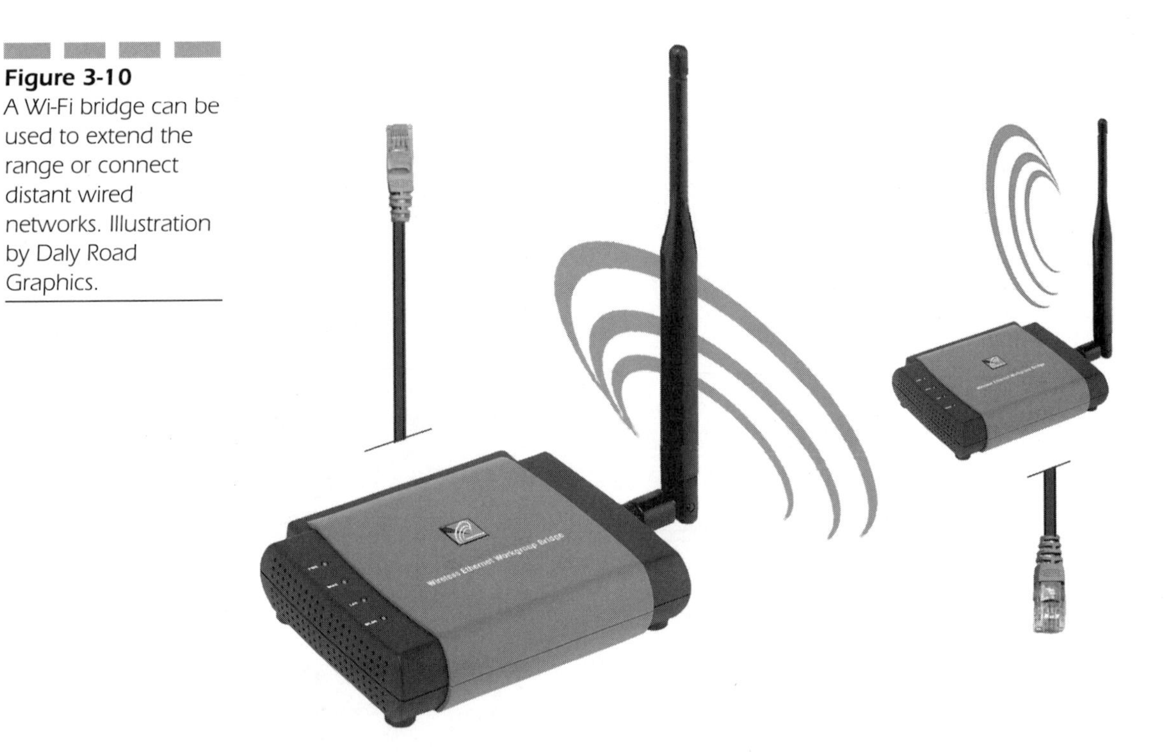

Figure 3-10
A Wi-Fi bridge can be used to extend the range or connect distant wired networks. Illustration by Daly Road Graphics.

range of a wireless network or fill in a dead spot. They can be used to connect a wired hub or switch or any Ethernet-compatible wired peripheral also, such as a printer or distant server. Together with an amplified transmitter and a focused, directional antenna, a network's straight-line range can be stretched for miles with these bridges.

Sources of Interference Another factor to consider, especially for base stations, is jamming. Wi-Fi shares the upper part of its 11-channel frequency band with microwave ovens. Modern ovens are well shielded, but don't ask for trouble by putting your access point on top of a microwave or even in the same room with one. Other sources of radio interference include older cordless phones, wireless video cameras, and some baby monitors. Check the label to see if they run in the 2.4 GHz band, and if they do, move them or the base station. It's not a good idea to plug the base station into the same outlet with one of these either, since the mutually interfering signals can leak through the power cords. Interference may not stop your link entirely or all the time, because it was designed to be robust. But noise on the air close to your Wi-Fi system's ear can force it to slow down. If you examine your configuration screen's signal report, and it shows a strong signal but slow throughput, you should turn off suspect sources until the problem clears.

We are going to cover security and how to prepare for it in the next chapter. (It is so important that it deserves plenty of detailed explanation.) But by this point, you should have plenty of choices as to how you are going to extend the Internet to your kitchen, basement, den, workbench, or backyard. The next chapter will guide you in configuring your Wi-Fi hardware so that it works efficiently and with privacy.

CHAPTER
4
Security
and Setup

Now you're knocking on my door. I hear you knocking, but you can't come in . . .

—Gale Storm

This chapter deals with armor-plating your *wireless fidelity* (Wi-Fi) home network. Radio waves can wander, making your system more vulnerable than you may have imagined, and many would like to take advantage of it. Electronic pests, such as spam, spyware, and viruses will appear on your electronic doorstep. This chapter covers threat assessments and provides suggestions to help you raise the drawbridge. Many of the most important security measures are installed when you first set up your network. We will cover that process too.

War Dialing

The movie *War Games* starring Matthew Broderick made its debut in 1983. A slang word from that film lives on today in the world of Wi-Fi. In that movie, a brash young man programs his PC to automatically dial a series of telephone numbers in a Silicon Valley area code and then compile a list of those that answered the calls with modem tones. Later on in the film, he fiddles his way into one of those numbers, starts a conversation with an intelligent military computer named Joshua, and nearly triggers a nuclear war.

The process of mechanically sifting through numbers is still emulated today by telemarketers and the annoying folk who send advertisements by printing them out on your fax machine. This is called *war dialing*. Unfortunately, as our telecommunications infrastructure has become more efficient, the threats and nuisances that come with it have also become more effective. Young Matthew's dialer could test targeted lines at the rate of three per minute. Today such invasions take place over the Internet where no dialing is necessary. Hackers can silently zoom through thousands of Internet addresses in an hour. Their computers do not need to sleep.

Dial-up modem users are connected to the Internet for just a few moments at a time. *Digital subscriber line* (DSL) users are now permanently attached to the Web, so they are permanently exposed. If you do not take steps to isolate your home net from these intruders, the odds that someone will grope your equipment rise from probably to certainly. Before we turned off incident logging on our home network, for example, we were getting an average of five warnings per day.

Unwired Security

Every networked computer shares a security risk. That has been true since the first network, and today's Wi-Fi nets are merely an extension of that. Wireless networking adds a new dimension because Wi-Fi has the defect of its virtues. The signal penetrates walls inward as well as outward. It is unlikely that you will ever see a stranger in your backyard attaching a cable across your lawn to his own house so that he could tap into your DSL access. The person or his handiwork would be noticed, challenged at once, and punished later. But someone peering in on your Wi-Fi home network will not be very noticeable and will be anonymous in any case. So in addition to the traditional threat coming into your system from the Web, a new problem exists with threats coming into your system out of thin air. It is no coincidence that almost all the experts interviewed in Chapter 8, "Where No One Has Roamed Before," spontaneously cited security as a major concern for new users.

It is true that some high-risk enterprises such as the Livermore Atomic Labs, nuclear power plants, oil refineries, and military installations have stopped using Wi-Fi networks altogether until foolproof security arrives. On the other hand, a growing number of California cities, including Glendale and Oakland, along with Orange and San Diego Counties, have embraced Wi-Fi technology as the wireless backbone of their civic networks. One advantage of wireless is that it is less vulnerable than wired networks in the event of an earthquake.

Perhaps you should look at it this way. If a skilled and determined burglar wanted to invade your home, no number of locks would keep him out indefinitely. But unless your last name is Gates or Rockefeller, your home would not merit that kind of time, attention, or risk. Network lockouts do discourage impulsive or curious behavior, particularly when security measures are combined or cascaded. Fortunately, protective measures are inexpensive and often free, except for the labor involved. Network security is part of the modern cost of living. It requires prudence, but not paranoia and certainly not panic.

Thanks to Wi-Fi's frequency agility, it is unlikely that someone could intercept your computer with a converted police scanner, as is the case with cell phones. The two *wireless local area network* (WLAN) techniques are called either *Direct Sequence Spread Spectrum* (DSSS) or *Frequency Hopping Spread Spectrum* (FHSS). With FHSS, a sender transmits a portion of data on a given frequency and waits for an acknowledgement from the receiver. If there is no acknowledgement after repeated attempts, both will

Hollywood Queen Patents Secure Communications (1942)

Hedy Lamarr enjoyed a successful pre-WWII career as a sexy movie star, first in Europe and later in America. But she possessed an agile mind as well as good looks. Her first husband (of six) was a domineering arms designer for the Nazis. From him, she understood the problem of securing communications to remote-controlled weapons. From him, she also learned how hard it is to abide with Nazis. Hedy drugged the maid he had hired to watch her, stole a servant's uniform, and fled from Austria to London. (see Figure 4-1). She later picked up her career in California, where she starred in many major films, opposite several glamorous actors (including Victor Mature in *Sampson and Delilah*).

When the Second World War began, she applied her inventiveness to helping the Allies. She realized that if a radio sender and receiver changed channels together, often but unpredictably, an eavesdropper would be unable to follow along. She had help from a music composer, George Antheil, who had experience in the coordination of sound and music scores. Mr. Antheil hit on the idea of using something like the punched paper rolls in a player piano to throw switches in a preprogrammed order for both ends of a radio link. Neither was technically sophisticated, but with a little help from engineers from the *Massachusetts Institute of Technology* (MIT), they submitted a proposal that resulted in a patent for their secret communication system. But the technology to make it work reliably didn't exist in 1942.

The U.S. Navy revived the idea, using transistors, to build a jamproof homing torpedo after their patent had expired. Lamarr and Antheil never received royalty payments, but they must have been proud when cell phone companies picked up their idea. As mobile phone callers move from one cell site to another, both the tower and the phone change frequencies in concert to maintain an uninterrupted connection.

In 1997, Lamarr was honored with an award at the Computers, Freedom, and Privacy Conference for "blazing new trails on the electronic frontier." Hedy Lamarr passed away in January 2000, but her idea lives on in every one of the millions of Wi-Fi computer interfaces in use today.

Figure 4-1
Technical creativity came from an unlikely source: the international film star Hedy Lamarr. In 1997, TRW, Lockheed Martin, and the U.S. Air Force gave her the Milstar award. They credited her for "paving the way for the successful development of the Milstar communications system." Photo courtesy of her son Anthony Loder.

switch to another frequency and try that. This minimizes interference, which typically sits on one fixed frequency.

In DSSS, a portion of outgoing data is mixed with a predetermined series of ones and zeroes called a *chipping code*. The extra data makes it possible to reconstruct a damaged portion of the data stream. Without these techniques, given wireless networking's low transmit power, its range would have to be extremely limited.

The latest twist to the technique is called *Orthogonal Frequency Division Multiplexing* (OFDM). It's basically the same, except that the sender and receiver determine which of many parallel channels are the clearest and preferentially push data onto the frequency slices with greatest clarity.

Frequency agility provides a basic level of privacy from fixed-frequency receivers, but another Wi-Fi-equipped computer can receive and understand the signal. Wi-Fi security is an ongoing problem that continues to be taken seriously. *Task Group* (TG) I is the security committee within the *Institute of Electrical and Electronics Engineers* (IEEE) 802.11 Working Group. They are currently working on drafting a text for a follow-up Wi-Fi security standard, IEEE 802.11i. It will include enhanced encryption formats and authentication mechanisms such as *Remote Access Dial-In User Service* (RADIUS), Kerberos, and IEEE 802.1x. Many of IEEE 802.11i's security enhancements will be possible through inexpensive *firmware* upgrades. Some will require completely new hardware. The research group In-Stat/MDR found that

- Enhanced encryption results in increased battery drain and slowness, which the standards group will have to consider as a factor.
- If end-user perception of security is favorable, the total residential and enterprise Wi-Fi and IEEE 802.11a unit shipments are expected to surpass 20 million by 2003.
- The fact that Wi-Fi products are enjoying much success in commercial markets makes it essential that security concerns are addressed, as data security is crucial in business environments.
- Networking equipment vendors as well as wireless specialists have been proactive in providing vendor-specific security enhancements to their WLAN solutions.

Even as things stand today, Wi-Fi provides padlocks that circumvent these eavesdroppers, but you must take the time and effort to use them.

Configuring Restricted Access

Some Wi-Fi system administrators cheerfully invite outsiders, such as the *freenets* described in Chapter 1, "Is This Trip Necessary?" Their express purpose is giving free Internet access to any passersby who might need it, setting their access points on a windowsill overlooking public areas such as city parks. They may advertise their presence with a chalk symbol on the sidewalk below. This practice is becoming common in Europe and has been given the name—you guessed it—*war chalking*. It may not be benevolent if a war driver tags an open system without the knowledge or permission of the owner. Historical precedents exist for these chalk tip-offs. During the Great Depression, wandering hobos did the same thing to point out householders who were likely to give handouts to strangers.

The generosity of freenets is possible because most router access points can dish out *Internet Protocol* (IP) addresses to and handle simultaneous use by as many as 64 clients. But unless these broadcasters have the permission of their *Internet service provider* (ISP), giving away something they have not paid for is a violation of their service contract. As of this writing, only one major carrier, Covad, and a host of smaller ones permit such giveaways. Others have abandoned unlimited usage privileges and charge according to traffic levels. Others, such as Time Warner Cable and AT&T Broadband, have threatened customers whose homebrew hotspots are advertised on freenet web sites. In some cases, ISPs have hired signal seekers of their own to ferret them out. As ISPs have seen, Wi-Fi again has the defect of its virtues. Any user with an access point can decide to piggyback other users on their connection—not only freenets, but also friends, neighbors, and small-time entrepreneurs who want to make their access points pay for themselves. It will be a continuing problem for service providers.

Some hotspot hosts are honestly and happily willing to pay the extra freight, such as coffee shop owners who use the feature to draw customers and keep them in their seats. In mid-2002, for example, the Fairmont Hotel chain announced that all 38 of its hotels in 6 countries would offer Wi-Fi to their guests.

War Walking A popular example of a hacker's cat-and-mouse game is called *war driving* or sometimes *stumbling*. Some stumblers have a friend drive them around while they watch their wireless laptop for a hotspot indicator. When they hit one, they pull over and try to log in. Two programs that can be used in this way are *NetStumbler* and its Apple cousin *MacStumbler*.

These hackers sometimes post their discoveries on the Internet without asking first. Those not old enough to drive have been known to strap a laptop onto the handlebars of a bicycle. Riding a bike while watching a laptop screen instead of the road seems like a certain way to land in the hospital, especially because screens are hard to read in direct sunlight. On the other hand, many hospitals are using Wi-Fi systems too, assuming that the laptop and rider survive the impact.

We conducted a test of our own with a laptop and our feet, war hiking up a fire trail overlooking our hometown. (We will call it Smallville, California.) Our primary interest (besides losing weight) was to judge the connectivity to our home system when given a line-of-sight perspective (see Figure 4-2). The *Global Positioning System* (GPS) told us that we were 500 feet above street level, the nearest home was 1/4 mile away, and our host system was 1 mile away.

That range to the house was too great for contact using our *Universal Serial Bus* (USB) Wi-Fi antenna as it turned out. But surprisingly, our laptop did sense several weak signals from access points that identified themselves as Smallville Fixed Wireless, even though no rooftop antennas were

Figure 4-2
A war driver sniffing the air. Photo by Daly Road Graphics.

visible. Finding systems and automatically logging onto them are such common tasks for Wi-Fi roamers that software facilities for doing so are standard issue from Wi-Fi hardware manufacturers. Free downloadable programs from other sources, such as boingo.com, make sniffing the airwaves, acquiring a Wi-Fi source, and automatically logging on as easy as a keystroke (see Figure 4-3).

A variety of other not-so-legitimate programs are available on the Web that perform these same functions and then go on to break the encryption used in thinly protected systems. As of this writing there have been two publicized uses of aircraft to sniff out and map wireless systems. One of these overflights was in western Australia and the other above Silicon Valley, south of San Francisco. In the latter case, war flying discovered many unprotected systems, even in the heart of this most technically sophisticated and security-conscious community. Their system administrators never imagined that a light plane quietly cruising overhead might have an antenna pointed downward at them.

Is War Driving Legal? The answer for now seems to depend on why you are doing it and what you do after finding a system. Simply searching for the presence of a signal and even attempting to automatically log on is legal. As previously mentioned, programs to do so are included for user convenience as part of every Wi-Fi hardware bundle. Presently, no laws forbid it, unlike the laws that make it a crime to listen to cell phone conversations. But if your intent is to vandalize someone else's data, steal his or her

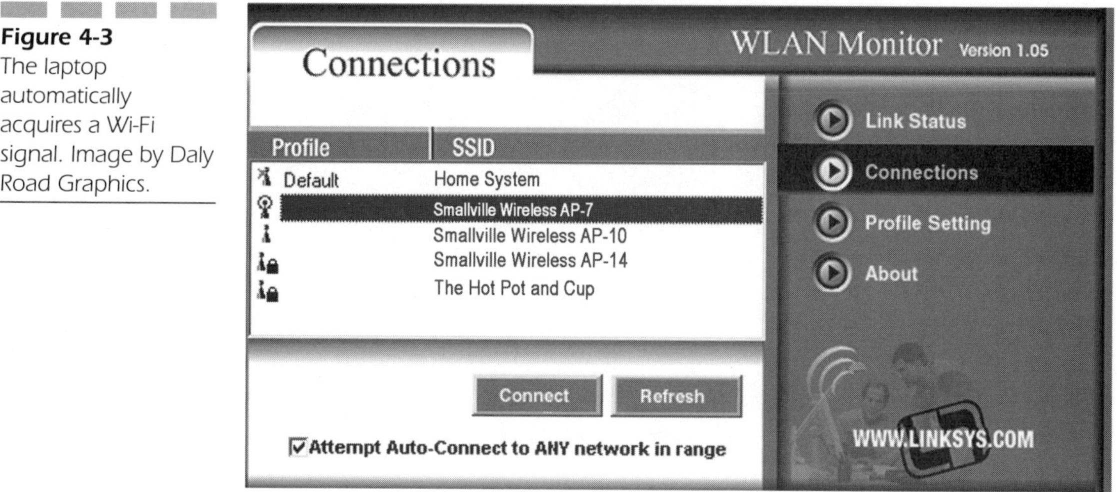

Figure 4-3 The laptop automatically acquires a Wi-Fi signal. Image by Daly Road Graphics.

Internet access, or any other malicious thing, then you may be liable in court. If such vandalism becomes a common problem, then legislation to punish it may follow. If you have to forge a password or decrypt a data stream to get in, then you have crossed the line of legality without question.

Prudent Wi-Fi network administrators will periodically do some war driving of their own. They want to determine the limits of their systems' coverage patterns and also to discover employees who have set up unauthorized access points for their personal convenience. These rogue office WLANs won't conform to the corporation's security guidelines and may broadcast a welcome mat for snoopers.

Some war drivers are organized so that they also put GPS receivers in their cars and continuously feed position reports into their computers along with signal strength readings. This enables them to draw detailed position maps of Wi-Fi coverage areas. Anyone who thinks his or her Wi-Fi system is unnoticed may be surprised to find it listed on a stumbler's web site along with hundreds of others.

Notes on Homebrew Antennas Many of the same manufacturers that sell Wi-Fi interface cards sell antennas for them also, with prices ranging from $60 to $250 (see Figure 4-4). Many cards do not come with a socket that will accept an antenna extension cable, so many hobbyists have resorted to modifying cards to add antenna jacks. Although the *Industrial, Scientific,*

Figure 4-4 This parabolic antenna connects Wi-Fi units up to 25 miles away. Photo by SMC.

and Medical (ISM) band is unlicensed and the *Federal Communications Commission* (FCC) doesn't require permits for transmitters under 100 milliwatts, the agency examines and approves all the equipment used for Wi-Fi transmissions. Carving on your card will not only void the warranty, but it will also change it beyond the FCC's recognition and approval. Even though the risk of getting caught is small, you probably don't need to do this because most of the equipment available from the factory is pretty efficient. Another one of the risks of hacking on your card is the risk of ruining it.

Even so, experimenters are making their own trick antennas. The most common type is known as a *yagi* antenna. These can be constructed from Pringles potato chip cans, PVC pipe, and one-pound coffee cans, among other things. Their purpose is to increase the Wi-Fi antenna's range by focusing radiated energy into a tight beam. Similarly, the antenna's receive sensitivity is boosted in the same way that a telescope increases the apparent image size by limiting the field of view. Although the raw materials are cheap, the designs can be complicated, and they must be built with precision to work well.

One reason for this activity is a perceived need to provide service to a distant area, such as a city park across the street from an access point. Another obvious motivation is an increased ability to snoop on systems with the stealth that comes from distance. We mention it in hopes that you will take the security requirements of your Wi-Fi system seriously. You may know your neighbors, but some of your biggest fans may be parked in a van ½ mile away.

Four Security Fences

In order to keep your data your own, four main tools are given to you as part of the Wi-Fi standard: the *service set identifier* (SSID), *Media Access Control* (MAC) address filtering, *Wired Equivalent Privacy* (WEP), and *virtual private networks* (VPNs). For peace of mind, you should use them all.

The SSID: Wi-Fi Networks Announce Themselves to the World

Leaving your house unlocked is not an invitation for strangers to enter. Leaving your access point open is. All Wi-Fi manufacturers include some software utility with their hardware that enables mobile users to search their vicinity for the presence of a hotspot as well as to report that system's SSID, its signal quality, and signal strength. Otherwise, how would they discover and connect to their own home system and adjust antennas for the best reception? One example of this function is Windows XP's *wireless zero configuration,* which automatically displays a discovered SSID in a list of available Wi-Fi networks. Free software can be downloaded from the Web that makes discovery and logon convenient for users of other network operating systems as well. All these tools can be used for mischief.

The SSID can be used to segment a large WLAN so that one department can be kept separate from another on the same floor. You can use it to keep your home Wi-Fi net segmented from other homes or curious passersby. Your Wi-Fi network's SSID was originally set at the factory. When you configure your access point and clients, you will have an opportunity to change it. If you do not make your SSID unique, you will share it with the networks that still bear the factory name. What's more, someone trying to guess it will have only a few possibilities to work through. Some preset SSIDs make it clear as to what brand of system they belong to. (Linksys uses the SSID "Linksys.") Some brands, like the Orinoco line, will automatically formulate a hard-to-fathom string of characters derived from the hardware's serial number. Otherwise, you are going to have to dream up one of your own. If you have the option, turn off the broadcasting of that name.

Authentication

Shared key authentication requires that a station attempting to join a wireless network must prove that it knows a secret key. Usually, that's the same WEP key used to encrypt data exchanges. First, the authenticator sends a challenge, to which the client returns an appropriate response. For it to work, all the wireless clients and the Wi-Fi router must use the same key.

Microsoft recommends *open systems* authentication, which skips the identification step. Their theory is that a hacker might eavesdrop on the introduction, extract the secret key, and quickly be able to decrypt everything else that follows. With open system authentication, a hacker can join your network, but won't be able to make sense of anything afterward.

MAC Address Filtering

Open system authentication, as it is called in the 802.11 specification, is not true authentication. The process relies on the identification of a station by its hardware address. Every Wi-Fi network interface has a unique serial number called a MAC address. It has nothing to do with Mac computers. It's usually written somewhere on the card (see Figure 4-5) or device. You can configure your Wi-Fi access point or router to allow connections only from specific Wi-Fi network cards—yours.

If you can't find the number written down, your PC can display the address. If you use Windows XP, 2000, NT, or Windows 9x, you first call up a (DOS) command-line window. Then enter the command *ipconfig/all*. If you prefer a graphical user interface, you can get the same MAC information by entering the command *winipcfg* (see Figure 6-4). Winipcfg is a handy utility that displays other information about your network interface as well, and it can also be used for basic troubleshooting.

On Mac OSx, open System Preferences, then Network, and then select the Airport from the drop-down menu. Its MAC address will be listed under Airport ID.

After you've listed all your MAC addresses, you can restrict access to them. This is sometimes called building an *access control list* (ACL). If any of your PC's addresses are not in the list, then your access point won't permit them into your home Wi-Fi network. Also, you should turn off your router's ability to automatically assign IP addresses to your clients, known as the *Dynamic Host Configuration Protocol* (DHCP). Since you won't have that many stations at home, it will only take a few moments to manually specify which MAC address will get which IP address. (It's a good idea to write these down as you go along.) You don't have to start at the beginning of the list of available addresses either, as a methodical hacker might. Your

Figure 4-5
A sample MAC, as shown on the underside of a PCMCIA Wi-Fi card. Photo by Daly Road Graphics.

router can draw from a large pool of temporary, nonroutable IP addresses, and you can choose from anywhere in the middle of that range.

The process isn't flawless. A sophisticated hacker can capture data frames from the airwaves and pick them apart to determine MAC addresses and clone them. Afterward, if your system is running in infrastructure mode, it will let them join. No method exists for MAC address filtering in ad hoc mode. Any hardware address can join an on-the-fly peer-to-peer wireless network.

Wired Equivalent Privacy (WEP)

In our experiment, Smallville's local fixed-point wireless ISP had wisely engaged WEP, a built-in Wi-Fi encryption feature, on all their access points except one. Our laptop obligingly picked the open one and connected to it (see Figure 4-6.) Our war-hiking experiment demonstrated the value of using WEP keys. We walked back home and promptly enabled this option on our own system (the default setting was disabled). The test also demonstrated the value of locating a Wi-Fi access point centrally in the household, so the signal covers desired rooms but not the entire block.

One important point to remember about WEP is that taken by itself, *it is not airtight*. A group of mathematicians in the Computer Sciences Department of the University of California at Berkeley discovered four methods

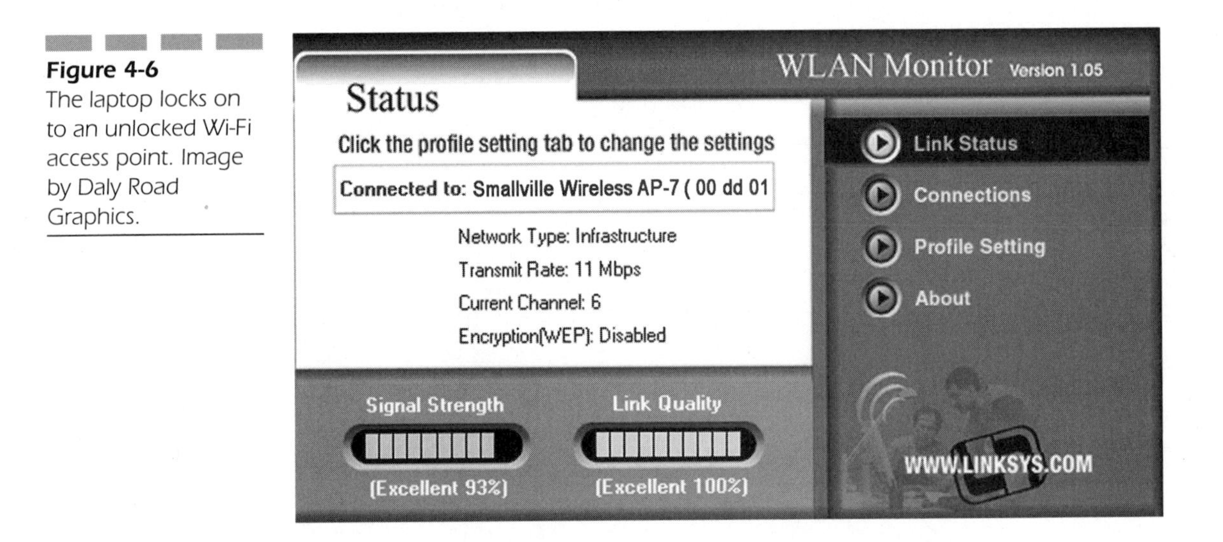

Figure 4-6 The laptop locks on to an unlocked Wi-Fi access point. Image by Daly Road Graphics.

for cracking it. They detailed each method on a public web page and concluded with this statement:

> Wired Equivalent Privacy (WEP) isn't. The protocol's problems are a result of misunderstanding of some cryptographic primitives and therefore combining them in insecure ways. These attacks point to the importance of inviting public review from people with expertise in cryptographic protocol design; had this been done, the problems stated here would have surely been avoided.
>
> —Nikita Borisov, Ian Goldberg, and David Wagner
> (reprinted with permission)

Another research paper by cryptographers Scott Fluhrer of Cisco Systems and Itsik Mantin and Adi Shamir of the Weizmann Institute also pointed out holes in WEP. Hackers used the Berkeley algorithm to write and publish Linux-based programs called *Airsnort* and *WEPcrack*. With a daylong sample, these can decipher WEP-encrypted traffic. The problem with WEP's RC4 encryption can be simply stated. Attached to each packet is the key that will enable the receiving device to unscramble it, but these keys are not dealt out randomly. They are dispensed in a fixed order. If a hacker can decipher one of them, then it is simple to predict those that will follow.

When you set up your Wi-Fi access point you will be given an opportunity to enter a WEP *key*. This is a unique string of characters used to randomize and to decode data as it is exchanged. Many systems enable you to enter a memorable password, which is then converted to a numerical, hexadecimal key for you. All the stations on your net must get the same treatment with the same key; oddballs will be excluded.

Keeping track of all these keys is a major problem for businesses that use Wi-Fi because they typically deal with many terminals, perhaps hundreds, that come and go frequently. Newer versions of WEP attempt to automate the process by dispensing keys from a special network server. The scale of your task is smaller, because you will likely only have a few stations to worry about.

WEP configuration gives you a choice of security levels, using 40-bit (sometimes called 64-bit), 128-bit, or, more lately, 256-bit encryption (see Figure 4-7). The higher the level chosen, the longer a snoop will have to sample your system to crack the key. Presently, that task may take a day or more, but higher levels mean more work for your PC on every data transfer, which may degrade your throughput. As previously mentioned, yet another WEP standard (802.11x) is in the works that will hopefully

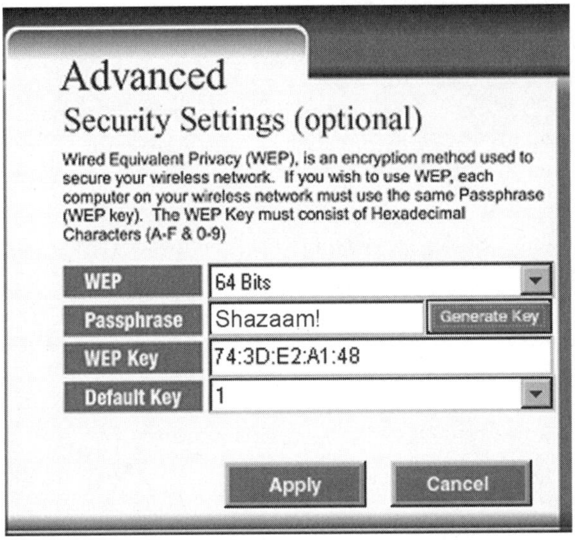

Figure 4-7
Setting up a WEP encryption key. Photo by Daly Road Graphics.

Advanced
Security Settings (optional)
Wired Equivalent Privacy (WEP), is an encryption method used to secure your wireless network. If you wish to use WEP, each computer on your wireless network must use the same Passphrase (WEP key). The WEP Key must consist of Hexadecimal Characters (A-F & 0-9)
WEP 64 Bits
Passphrase Shazaam! Generate Key
WEP Key 74:3D:E2:A1:48
Default Key 1
Apply Cancel

keep honest users one step ahead of the hackers. As of mid-2002, it is still being revised. Computer scientist William A. Arbaugh of the University of Maryland and grad student Arunesh Mishra found two weaknesses in the IEEE's first effort, which they named *man in the middle* and *session hijacking*. The IEEE's TG I is working on those problems.

You should check the web page of your equipment manufacturer from time to time to see if new security enhancements are available for download. Be sure to change your WEP keys periodically.

Virtual Private Networks (VPNs)

The number of people who work from their homes is increasing, and one of the reasons is that corporations are closing down rented remote facilities to save money. Of course, many employees willingly work from home as well. In any case, they must send information to and from their corporate offices, and do so securely. But unlike a corporate intranet, home office workers' data must traverse insecure links, including their Wi-Fi home network. (For a detailed explanation of how VPN works, see Chapter 7, "Virtual Private

Networks.") Briefly, VPN provides a secure transport by encrypting data and putting it inside an unencrypted IP packet, or envelope. These envelope packets cross the Internet like any others. But when they arrive at the corporate intranet, their payload is stipped out, decrypted, and retransmitted as though it had originated on the headquarters campus. Outgoing data gets the same treatment when it arrives at a company PC in a home or apartment.

Implementing VPN can be done in more than one way, so if you must use it at all, you will have to get advice, assistance, and probably software from your company's *Information Technology* (IT) department. It's not something you can do on your own, if for no other reason than you won't be able to get into your company's intranet in the first place without logins or passwords. (If you can, then something is terribly wrong.) You will not need VPN for noncommercial endeavors. Some residential routers will not support it, so check before you buy. You will need a static IP address from your ISP to make it work.

Password Advice

Modern-day citizens are inundated with passwords. We need them for ATM machines, door locks, mail systems, and so on. It is bothersome to keep track of them all. Having to change them at times just makes it worse. It helps if you make password renewal part of a regularly scheduled maintenance routine (see Chapter 5, "Good Housekeeping"). The most common passwords in use today are *sex* and *password*. The word *public* is also a bad choice, because industrial-grade network hardware is usually shipped with that default management password, which is sometimes called a *community string*. You can do better than any of those choices by picking one that is

- At least eight characters.
- Not a phrase or word that someone else would recognize as such.
- Not something prominent about yourself, such as your dog or spouse's name.
- Not following a pattern, such as the day of the week plus the date.
- Partly numerical. Add the year you graduated or something similar.

Also, a good password is something that you can remember without writing down. Many people do just that, however, and then put the yellow sticky tag on their display screen for the world to see. Hundreds of password-cracking programs are available on the Net. If an exact match to your password can be found in the dictionary, then it's not much of a password.

Configure from Inside

As mentioned, one of the nifty features of Wi-Fi access points/routers/firewalls is your ability to set them up as you like from any Wi-Fi client using a web browser. You may also do so from a wired client and even from the Internet, if you think you must. But the feature is very convenient for hackers also. Turn it off if you can, and do your configuration from a local hardwired terminal. Again, change the manufacturer's preset logon name and password for your access point (see Figure 4-8.) If you don't close that door by changing these, anybody can log in from the street or the Internet, seize control of the access point, and grant themselves godlike privileges. They could even lock you out of your own network. To get back in, you would have to reset the device, wiping its memory and then reconfiguring it from the start.

At configuration time, some brands of access points give you the choice of running an *open* or *closed* system. This refers to the names given to clients. If your neighbor has a computer named "any" or is completely nameless, then he will not be able to jump into your data pool if your system is set to closed.

Internet Addressing Basics Telephones would not be of much use without identifying numbers. Phone numbers not only allow one phone to be picked out of millions of others, but through its area code, the number provides information for locating and connecting a single phone within the sys-

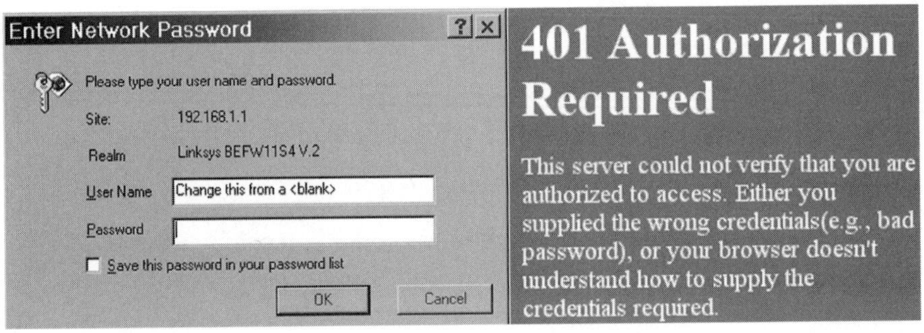

Figure 4-8 Change your router's password to give strangers a cold shoulder. From Daly Road Graphics.

tem. Every device on the Internet must also be identified with a unique number or *IP address*. Like a collection of zip codes, this system of numerical identification tells automatic equipment, called *routers*, the best way to set up a connection between two or more stations. Unlike permanent phone numbers or street addresses, most routers can be called upon to temporarily assign, or *lease* IP addresses to client stations requesting them. Routers do this in a process known as *Dynamic Host Configuration Protocol* (DHCP). They can also download other useful client information, such as an IP subnet mask and the addresses of domain name servers. Home Wi-Fi routers do this also.

You may ask: Why do addresses have to be temporary? In fact, most Internet addresses permanently refer traffic to one site. But when the system was first designed, nobody anticipated millions and millions of users jumping onto it so quickly. The amount of available numbers is getting scarce, so when you rent one today, the law of supply and demand dictates that you will have to pay for it. A fee of $10 per month is typical. The next generation of IP, known as *IP version 6* (IPv6), will handle many more addresses. It will accommodate colonies on other planets in fact, but it will also require a complete overhaul of the infrastructure we use today.

Users today, particularly those dialing in, only need an address for the duration of their session. That address can be reassigned when they are through with it. One common misconception about dynamic IP addresses is that they are somehow safer than permanent ones. A short-life IP only makes it harder for a hacker to find you in particular or find you a second time.

Unique Network Address Translation (NAT) Parameters Most Wi-Fi access points can do *Network Address Translation* (NAT) using configuration values preset by the manufacturer, known as *defaults*. Choosing defaults is simpler and easier than choosing values for yourself. But hackers will know those default values also, which gives them a starting point for their puzzle solving. Don't make it easy for them. If the game is difficult from the start, they will likely go somewhere else to play. For example, if you only have five PCs that you expect to use on your home Wi-Fi network, tell the router to provide no more than five addresses period.

Manufacturers want to simplify the setup process and make their equipment fast and easy to use. Many have done an admirable job in that regard. But if security is a concern for you, as it should be, then you cannot take what you are handed and leave it at that (see Chapter 8's interview with Jim Harrington.) After you get your system working, you will have to read the manual and then go back to change passwords and parameters from their defaults. At prudent intervals, you may have to change them again. Consider it part of the cost of digital living.

Workstation File Sharing

A major benefit of Wi-Fi home networking is the ability to easily share files and programs among several PCs, thus multiplying the amount of data available to all. Windows 95/98/Me has this feature built in. It is tempting to use file sharing to make everything available from anywhere at once and be done with it. But if you take the easy way, you may wind up sharing your financial information, credit card and social security numbers, and heaven knows what else. The same can be said if you share items individually, but use simple passwords or no passwords at all.

You can take two measures to minimize this risk. First, throw out all the network protocols you don't actually use. This will probably leave you with only the *Transmission Control Protocol* (TCP)/IP. Earlier versions of Windows include *NetBios Enhanced User Interface* (Netbeui) and *Internetwork Packet Exchange* (IPX) and have them turned on by default. Second, you should remember that you only need Microsoft networking for sharing files and printers with other Microsoft PCs. If you don't need to share locally, you will not need Microsoft networking at all to browse the Web. A reasonable compromise is to share only those folders that you will need to share. One way is to create a folder named Transfer on the root directory of your PCs. It functions like a temporary mailbox, holding files that are due to be swapped or that have just been received. Only the files in that folder are at risk and only for as long as you leave them there.

Firewall Limits

Most new users are surprised to find out what a free-for-all the Internet has become. Firewalls are devices that analyze and selectively block traffic between your Wi-Fi network and the outside world. Some firewall functions are probably built into your Wi-Fi access point already. They provide a barrier to help keep intruders from infiltrating your network through your router. But the router firmware is not clever enough to analyze data that you invite in, such as email messages. If you pick up a broadcast virus, for example, which automatically sends copies of itself to everyone on your personal address lists, then your firewall won't stop it from going out. You can also pick up a virus from an infected floppy. Your router's built-in firewall can't monitor those kinds of viruses or help eliminate them once they've arrived.

You will gain additional protection against these *Trojan horse* programs and email viruses by adding software firewalls to each of your home network PCs. Fortunately, some of these are available for free, such as ZoneAlarm, from zonelabs.com. Others such as the Tiny Personal Firewall from tinysoftware.com, the Norton Personal Firewall from symantec.com, and mcafee.com's Personal Firewall sell from $30 to $60. As with virus detectors, these programs need periodic updates to keep up with the latest tricks. As virus writers become more sophisticated, more viruses will escape detection in any case. Most antivirus utilities will automatically cease updates after a year unless you buy another year's worth.

One advantage of using a personal software firewall is that it runs on the PC you are using, so you can actually see how many and what kinds of attacks are coming at you as they occur. This can be an enlightening experience, as you learn that supposedly legitimate web sites are actually trying to plant spyware on your PC without your knowledge, that a good friend has unknowingly sent you an email virus, or that an anonymous cracker is electronically rattling your PC's doors to see which of them can be opened. One disadvantage is that it can be a nuisance when it does what it's supposed to do. You may discover that you cannot connect to some resources on your own home network unless you specifically and individually permit it.

This personal firewall software is so popular that Windows XP includes an *Internet connection firewall* (ICF) as a built-in feature. If your network uses Microsoft's *Internet Connection Sharing* (ICS) to share Internet access among several home computers, they recommend that you use ICF on the shared Internet connection. The ICS and ICF services can be enabled independently. Microsoft advises against enabling their firewall software under the following conditions:

- If the connection does not directly connect to the Internet
- If your network already has a firewall
- If you use a VPN link
- If the PC is a network client, because ICF will interfere with file and printer sharing

Here's how to turn it on:

1. Click Start, click Control Panel, and then double-click Network Connections. Click the dial-up, LAN, or high-speed Internet connection you want to protect.

2. Under Network Tasks, click Change settings of this connection.
3. Select Internet Connection Firewall and then the Advanced tab.
4. Check the box Protect my computer and network by limiting or preventing access to this computer from the Internet.

Firewall Fire Drills

Another way to search for holes in your Wi-Fi Network's shielding is to deliberately ask a special-purpose automated system to attempt a break-in. Several web sites will on request probe your system and report weaknesses. These services perform TCP *port scans* and also check for common mistakes like the open file sharing mentioned previously. During a port scan, a piece of software sends TCP/IP interrogations that generate a response if that port is open or "listening." Each type of IP traffic has to go through a certain port number, so some should be open or you would not be able to receive information from the Internet at all. The following table shows IP ports that normally exist, the programs that use those ports, and how they are used. You don't have to worry if you see these listed in a security report about your system.

IP Port Number	Program Name	Purpose
7	Echo	Pinging
21	FTP	Copying files
23	Telnet	Login to remote terminals
25	SMTP	Email
53	DNS	Name—IP address resolution
79	Finger	Station identification
80	HTTP	Web browsing
110	POP3	Email
119	NNTP	USENET news reporting
162	SNMP	Remote control of network devices

Some programs, including ones you never asked for, will hold ports open continuously. Ports are also holes that hackers can exploit to drill into your computer network, so the less ports open, the better.

The following examples of automated testing sites are representative of the self-exams you can find on the Internet. They all require you to enter your email address and log in. They would all like you to subscribe to more thorough, permanent, and expensive examination service plans.

- **Shields Up at grc.com**
- **HackerWhacker™ at hackerwhacker.com** The user is automatically emailed instructions on how to get to the tests. You will have to know your access point's IP address. Look for smiley faces. If you get a frowny face, you have a problem.
- **Security Space Desktop Audit at secure1.securityspace.com/sspace** The basic and single-use tests are free.
- **Vulnerabilities.org** You are referred to the ScannerX web site and are asked for an email address.

If you pay for a super-thorough test, it may take hours to complete. Even so, asking a server for an exam is much better than having it done for you by a malicious human. If the tests show that you are okay, you have gained some well-earned peace of mind.

Unwanted Bugs: Viruses

Computer viruses are a fact of life. Everyone, lazy or diligent, gets them sooner or later. Wi-Fi networks provide an efficient method for distributing them among all your PCs at home, at a corporate office, or in between. Here are some real-life virus scenarios:

- You open an email from one of your coworkers labeled "baby pictures." By the time you realize that the text makes no sense, you notice that the hard drive activity light is on solid. A virus is charging through your files like a bull in a china shop.
- Three months after disinfecting your PC of a virus, you put in a floppy disk that you used while it was infected. It's baaaaack!
- You buy a junk PC so you can salvage its hard drive. After you add it to your own PC, instead of formatting the new drive at once, you check to see what's on it. Surprise!
- A buddy at work gives you a copy of his "free" recipes, screensavers, games, or whatever. It's been copied and recopied from a dozen PCs. You get a free bug.

- Before your PC comes back from the shop, the technicians conscientiously run a test program on it to guarantee that it's fixed. They use the same floppy (not write protected) on every PC before they ship it. No extra charge for the Trojan horse.

It's possible to write an entire book on the subject of viruses alone, but here are some basics. As of mid-2002, there were over 50,000 of them, in a large variety of insidious, devilish forms. (An extensive encyclopedia is available at symantec.com.) Some will sleep for months before announcing themselves. Some will hide by changing shape. Some commandeer your mailing list and mail copies of themselves to everyone you know, signing themselves with your name in the hope that your reputation will get them through the target's door. If a virus gets loose in a large organization, the lost productivity plus the man-hour cost of cleaning it out can be staggering.

Keeping up with new viruses is a full-time job. Several software companies will do this for you. McAfee and Symantec are two of the most popular examples. In addition to purchasing their software, you must periodically connect to their web sites to download new virus *definitions*. These are numerical descriptions that enable the software to pick a virus from the thousands of uninfected files on your PC. Hundreds of new definitions are added every week. Virus software performs multiple functions:

- It scans your hard drive and memory in search of infected files.
- It attempts to restore an infected file to its original state.
- Failing that, it will quarantine the file so that the virus cannot reproduce.
- Optionally, it will delete the infected file, which you may have to replace from a backup.
- It stops viruses on the way in.

The last function is probably the most important. The program examines all data as it enters your PC from whatever media is used in order to flag and stop infected files before they can do damage (see Figure 4-9). This gives you the opportunity to call and warn the owner of the source files. An immediate response to a virus warning is crucial to preventing epidemics. Usually, you cannot be blamed for getting a virus. Unfortunately, your friends, associates, customers, and other home Wi-Fi users will blame you for spreading one.

In addition to the previous software, a free scanner is available at housecall.antivirus.com. Unfortunately, their scanner won't find a virus that hasn't hatched yet. It finds active infections.

Figure 4-9
Another virus arrives.

Security Alert Details

Logged Event: Default Block DeepThroat Trojan

OK

Spyware

The Internet provides valuable resources, but it is often used to gather information about you without consent. Some businesses make a profit by compiling dossiers, known as profiles, which store your surfing paths, buying preferences, employment information, medical data, credit card numbers, and other surprisingly personal information. This is called *data mining*. What's worse, after they have it, they sell it to others without your knowledge, much less a guarantee that it won't be used to your disadvantage.

Sometimes this happens when you are asked to render information, which is then stored and never forgotten. One example of this is the submission of an online job application or resume. Web site submittal forms may be presented to you with a permission box prechecked. At some sites, you may have to specify that you don't want to receive advertisements or to have your choices cataloged, but those options are buried on a back page. In any case, you have no guarantee that your preferences will be honored.

Someone Is Watching You *Cookies* are small text files that are stored by your browser as you surf from web site to web site. Some of these enable you to log into a site once and then return without having to log in again. Most surfers are unaware that as they jump from site to site, these cookies are accumulated and used to silently compile a trail of surfers' wanderings. The largest of these tracking companies is *DoubleClick*, which came under investigation by the Federal Trade Commission and others for some of its practices. (See "Spyware Makes Headlines"). They use cookies to track your movements about the World Wide Web. Their computers are patiently watchful and never forget.

Browsers can be configured to reject cookies or warn you when they are being sent, but some of these "Kilroy was here" files are actually helpful. You may not have to keep logging into a familiar web site if you have a cookie from it that identifies you. But you will receive so many cookies that you will most likely tire of selectively examining all to stop a few.

Spyware Makes Headlines, August, 2002. New York's Attorney General Elliot Spitzer announces a settlement with the nation's leading Internet advertising service: "It's hard for consumers to trust e-commerce when they can't see the practices behind the promises," Spitzer said.

DoubleClick collects consumer data while it displays web-page banner ads and provides other e-commerce technology services as a contractor to websites. Through its widespread network of website clients, DoubleClick is able to use its cookies to track the surfing activity of any given computer.

New York led the 10-state, 30-month inquiry into DoubleClick's privacy practices. The investigation focused on how DoubleClick discloses its practice of assigning anonymous but unique cookie identifiers to the computers of websurfing consumers.

DoubleClick ignited controversy in early 2000 by announcing that it would use personally identified profiles to bolster the appeal of online banner ads. State attorneys general, private litigants, and the Federal Trade Commission focused on the company. DoubleClick soon scuttled the plan to use personally identified information and the Federal Trade Commission closed its investigation. In May of 2002 DoubleClick concluded a related consolidated class action by agreeing to adopt privacy disclosure standards and make a payment of $1.8 million in attorneys' fees.

"When an online contractor can invisibly track nearly every online consumer, consumers deserve to know the privacy cost of surfing the Web," Spitzer said.

The New York State Attorney General's office, August 20002 http://www.oag.state.ny.us/press/2002/aug/aug26a_02.html

A surprising amount of information can be gleaned from your browser every time you pull up a web site, such as the URL of the *previous* web site you visited. The firewalls described previously help to keep a lot of his *Trojan horse* software out, but not all. Some of it actually rides in at your request, along with free software that purports to speed up your downloads, or with free entertainment players or media.

Industrial networks have teams of specialists armed with expensive hardware and software to protect employees from outside threats. Your home Wi-Fi network is just as vulnerable as the one at work. In some ways, it is more vulnerable because you use it differently.

For a typical, real-life example, I rented a DVD and enjoyed it on a Wi-Fi laptop out on our patio. As the movie began, the DVD asked if we would like some free software that would enable screen captures and do other nifty tricks. On invitation, it loaded itself over the Wi-Fi network link and then the movie played with no problem. But from that point on, every time a DVD was played, the PC secretly used the Wi-Fi Internet link to report which movie was being viewed, plus a viewer identifier, to an unknown host. Any network-attached PC would have been as vulnerable; it just was not obvious with a free-floating laptop. The deception only become apparent after firewall software had been installed. It took some time to remove the spyware DVD player and recover the harmless one that originally came with our laptop (see the Chapter 5 section, "Call for Backup").

Those annoying pop-up banner ads can also become snitches by using buried *web bugs* or invisible *Graphics Interchange Format* (GIF) files. Known as *ad bots* or *ad ware*, these programs can welcome ads onto your computer even when the program they rode in on is not in use. This spyware can slow your PC and eat up memory, disk space, and processor power. It is prudent to periodically scan for and remove it. Some of the free programs that have been infiltrated by this surreptitious software will stop working if it is removed. A few will even crash your PC in protest. But by regularly tossing out the digital garbage, you enhance the security of your home as well as your PCs, because some spyware is designed to garner specific personal information about your household.

Spy Versus Spyware Some defensive programs come as part of bundled packages of utility software. These combine the functions of antivirus, firewall, privacy guard, and parental control programs. Bundles are cheaper than buying the programs separately and sometimes more efficient as well, since they can share common databases and subscription downloads.

The cat-and-mouse game between predatory content providers and their victims is a nonending siege, just like the contest between virus authors and their victims. New threats occur every week, so protective programs must periodically be supplied with updates in order to recognize new threats. Providers of shield software naturally use the Internet to distribute these new files, which makes the process simple or even completely automated. As with virus and firewall subscriptions, the service will expire, typically after

Here is partial output from a spyware extractor program, showing a few of the files surreptitiously planted on our home PC. In time, these can accumulate into hundreds.

Suspicious Registry Keys:

Alexa key: HKEY_LOCAL_MACHINE\software\microsoft\internet explorer\extensions\{c95fe080-8f5d-11d2-a20b-00aa003c157a}\

Network Essentials: C:\PROGRA~1\IDT\NET2FONE\net2fone.exe

Deleting: C:\WINDOWS\Cookies\Wi-FiBench@**servedby.advertising**.txt

Deleting: C:\WINDOWS\Cookies\Wi-FiBench@**fastclick**.txt

Deleting: C:\WINDOWS\Cookies\Wi-FiBench@**doubleclick**.txt

Deleting: C:\WINDOWS\Cookies\Wi-FiBench@**valueclick**.txt

a year's time. At that point, you must renew the subscription for a price, buy a newer version of the package, or fall behind.

Other low-cost alternatives are as follows:

- **Ad-aware by www.lavasoft.nu** Free! And free subscriptions! It automatically searches and removes spyware.
- **Spychecker at www.spychecker.com** This scans downloaded software on the way in.
- **WebWasher by webwasher.com** Free to home users. It filters out banner ads, cookies, and other web junk on the fly.
- **Proxomitron at www.proxomitron.cjb.net** This filters out http header files, GIFs, and other snoops.
- **CyberScrub by cyberscrub.com** Free 30-day trial, $50 afterward. This scans your hard drive and obliterates deleted files, emails, and cookies, and wipes swap files.

Another way to protect yourself is to do your surfing through an anonymous translator service. After you subscribe and log onto these, they go out to the Web and pass information back to your browser all the while presenting a blank face to the sought-after web pages. You can get a free trial version of one of these at www.anonymizer.com.

Spam

We are not referring to the processed meat that comes in cans, but rather the blizzard of junk mail that soaks up your time and costs ISPs so much to (partly) filter out. Although it can be used to deliver email viruses, spam is not a severe security risk other than to notify an Internet predator that you are available as a target. As with the false-front web browsers previously mentioned, blind proxy or relay services are also available for email.

Your best spam defense is to refuse to give out your email address unless it's necessary. Incidentally, if you receive an offer from a spammer to get yourself taken off their mailing list, don't answer. It's just a ruse to guarantee to them that you really exist. You'll wind up on their premium target list instead.

Years ago, many software companies would provide free samples of their wares. Today many of the samples are without money charges, but may be passed out solely to obtain the fact of your existence and your email address. A program advertised as free often will require you to enter a working email address, which is then used to provide you with an activation code by return mail. The mail message then asks you for a credit card number, which will be used to permanently activate the program, which will otherwise go dead after a few weeks. This is not exactly the same as free.

Most ISPs will give you several email addresses if you ask. You can devote one of them, joes_trash@provider.net, to low-priority mail. You can use this dummy account to accumulate spam and erase it all at once, or use it to track how many spam sources originated with one submittal. You can also periodically delete the account and create another spam trash bin. This means the spammers will waste time and money sending their junk to nonexistent addresses.

It is important to remember that information flows in only one direction: out. If you give up your email address, credit card number, or anything else of value to a faceless entity on the Net, you will never be able to call it back or even guess to whom it will migrate.

Kid-Safe

One special Wi-Fi problem for concerned parents is their unwired laptops, because these tend to wander about the house or, more recently, to and from school. You cannot always isolate them simply by locking a door. This increases the possibility that curious youngsters may use them to find Internet sources that aren't good for them to see. One classic example is the web site whitehouse.com. Kids doing a school report on the president's home (whitehouse.gov) might hit the dot-com page instead and be treated to pictures of naked women. Countless other similar sites exist, too many to list. Some are scary or dripping with hatred, or provide information on how to do truly dangerous mischief. New ones pop up constantly.

The second-best way to protect kids on the Internet is to install parental control software on your mobile Wi-Fi PCs (see Figure 4-10). Dozens of programs, such as Cybersitter and McAfee's Internet Guard Dog can be installed to censor what comes on your PC. In addition to forbidden web sites, they monitor traffic from newsgroups, chat rooms, and even email, zapping nasty words on the fly. Most censors will log attempts to reach banned sites and compile reports, and some will even email an alarm to parents when it happens. Parents will have to customize blocking if they have the option, so that the Internet will remain useful to older kids. Presumably, elders will turn the censor off for their own use, but you should

Figure 4-10
Parental control software helps keep your Wi-Fi laptops muzzled. Photo by Daly Road Graphics.

check for a timeout feature when you buy that will turn blocking on if a PC sits unattended for a specified period of time.

The best way to protect kids in the cyberage remains the old way, which is to be aware of what your children are looking at, just as you do with TV and magazines.

Walking Laptops

The term SOHO was invented to describe a new phenomenon that is driving the growth of Wi-Fi home networks. More and more people are working from the *Small or Home Office*. The U.S. Department of Labor's mid-2001 survey showed that corporate employees are carrying their work home, either for a few nights a week or full time, as do owners of small businesses that operate from their homes. Over 15 percent of the workforce, that's 20 million people, stated that they "usually work from home." The trend is growing as corporations downsize or shut down remote rented facilities.

In August 2002, Iowa senator Chuck Grassley reacted to an internal audit by the Treasury Inspector General for Tax Administration.

"In recent days we have seen Inspectors Generals' reports of the Federal Bureau of Investigation (FBI) and Customs Service having thousands of computers that are lost, stolen or missing. Earlier in the summer we learned in another Inspector General report that the Internal Revenue Service (IRS) had approximately 2,300 computers that were lost or stolen. I'm worried that just as dryers have the knack of making socks disappear the federal government has discovered a core competency of losing computers," the senator said in a letter to Mitch Daniels of the *Office of Management and Budget* (OMB).

The missing computers could contain private taxpayer data. "Information on tax forms is regarded as a prime target for identity thieves," the audit said. Other Department of Justice agencies that reported a loss of computers and weapons included the Bureau of Prisons, the Immigration and Naturalization Service, the United States Marshals Service, and even the FBI.

The chief tool of these wandering workers is the laptop computer, because like Wi-Fi networks, they provide portable computing power and a means of picking up and transporting huge data files. Now that most laptops come pre-equipped with Wi-Fi interfaces, it makes more sense to have a Wi-Fi home network. A Wi-Fi SOHO facilitates the transfer of data picked up at customer sites and corporate offices into more powerful desktop computers located at home, with their greater storage, comfortable keyboards, and larger visual displays. After data is processed, a laptop is an ideal way to carry it back to a customer site. So laptops not only migrate from room to room in a Wi-Fi residence, but also out the door to the world at large. Inevitably, some personal data that might never be put onto the Internet migrates onto laptops and is carried about as well.

Not only is a laptop easy to carry, but it is also easy to carry off and resell. The most valuable part of it is often not the hardware, but the data on the hard drive: personal data ideal for identity theft, such as credit card and social security numbers and bank correspondence. It also can include corporate data such as customer lists, sales strategies, internal phone numbers, new product research, and so on. Information that is usually kept under lock and key in a corporate office is often hauled about in a laptop with no protection. It's not surprising, then, that a new breed of specialized criminal has evolved to specifically target laptops, particularly in airports.

The problem is huge. The Computer Security Institute is a trade organization made up of computer and network security professionals. They estimate that 57 percent of firms suffered losses from laptop theft in 1999. A laptop insurance company, safeware.com, says that in 2001, over 590,000 laptops were stolen. Only virus attacks were a more widespread security problem.

You can take action to protect yourself. Obviously, a variety of locks and chains for laptops are on the market, but they make mobile laptops less mobile for the good guys as well as the bad guys. Some modern laptop PCs are sold with built-in radio beaconing devices, the PC equivalent to the Lojack homing system used to recover stolen cars. Some services offer automated methods to back up the contents of a PC (which is easy to do on your own if it is Wi-Fi capable). These measures may ease the sting of a loss, but they will not retrieve confidential data.

One example of a proactive protection is the *pretty good privacy* (PGP) program from www.pgp.com. It is used to encrypt individual files, especially those you plan to email or wirelessly copy onto another PC in a public hotspot. BcCrypt, from www.jetico.com, goes further. It sets up a virtual hard drive on your laptop, one whose data is encrypted. Windows XP offers hard drive encryption, named *Encrypting File System* (EFS), as a built-in feature. With it, application programs can read and write to this virtual drive, but a thief looking to ransom your corporate secrets will see nothing without the password.

Voluntary Amnesia

Laptop crooks are not the only cyberpredators who want to know all about you. Some ISPs (mostly cable companies, but not exclusively) don't even have privacy agreements. They've been known to sell detailed information from their user databases, including names, addresses, and lists of URLs visited. In 2002, Michigan's Comcast was sued for this kind of activity. Lawyers accused them of recording the browsing activities of a million subscribers. For their part, Comcast attorneys responded by asserting that the data in question did not contain "user-identifying" information.

You may be interested to know that your web browser and your PC retain many details about whatever you look at. Part of this takes place in order to increase your system's efficiency. When you download a web page picture of any kind, it is assigned a unique identifier and stored in a *cache* for future reference. Afterward, if you revisit the site, the cache is checked to determine if the graphic is already available locally. This may have been necessary in the age of slow telephone modems, but with broadband links it makes little difference in the time taken to paint a web page on your screen. Web browsers also maintain a history of the sites you've visited, so that you can go back to them by typing the first few letters of their URLs. This data will accumulate over time and can give a detailed picture of what you do as well as eat up your disk space. Browsers have facilities for periodically erasing much of it, and if you take the time, you can configure them to store little of it. But Browser cleanups are not thorough. Even erased files can be brought back to light by a knowledgeable hacker who can also be an effective tracker. Utility programs that unerase files or recover damaged files are commonly available to users who have accidentally erased something. But these same programs, or more sophisticated versions of them, can be used to restore data that was deliberately disposed of.

This problem becomes more acute if you work from home. Many companies require employees to flush their email bins from time to time and observe other security precautions. But workers in a home environment are more likely to feel deceptively safe behind their own walls. You may have need of a file shredder program if you plan to resell your PC or hard drive.

Fortunately, many different utility programs are available to deal with these problems. Some can be used for evaluation before you have to pay. Some are really and truly free. They will perform the following tasks:

- Zap Internet temp directories, history files, and the index.dat file.
- Let you decide which of the cookies you want to keep.
- Give you a hot key that will instantly replace your browser display with something uninteresting, should you be approached from behind (sometimes called a *boss-coming key*).
- Stop some clever web sites from altering the way your browser behaves.
- Erase stored autocomplete site addresses.
- Erase your recent files lists.
- Dump the recycle bin.
- Erase browser favorites.
- Erase anything else you specify.
- Run as often as you determine.

Here are some specific examples (*not* endorsements): Historykill from swanksoft.com and Internet eraser from interneteraser.com are written specifically to deal with web wanderings. SurfPal and PopupStopper by panicware.com make browsing more enjoyable, if not safer, by suppressing annoying pop-up windows. BCWipe by jetico.com focuses on the problem of deleted files, unused sectors, and swap space. (They aren't really deleted until they are repeatedly overwritten.)

Setting Up Your Network

Manufacturers have spent considerable effort through three or more generations of Wi-Fi equipment design so that you will not have to spend much effort to make it safe and work correctly. New advances such as Universal Plug and Play promise to make setup even easier as the terminology, marketing, and methodology become more standardized across brand names. In any case, if you've had any experience installing PC peripherals, this process should be even easier. Usually, you won't even need tools other than a keyboard and an instruction manual. You may need a little advice as well, but that's why this book was written.

Pulling the Wrappers

Your first temptation will be to pull the cellophane off of all the parts to see what you've purchased. Turn the box over and around first to see if it has already been opened. Sometimes merchants will reseal returned equipment and put it back on the shelves. Sometimes users in search of spare parts will get them from a package and walk out with them in a coat pocket. That doesn't happen often, fortunately, and the best time to check for evidence of it is when you first pull the box from the store shelf. Looking first is a good habit to cultivate. *Let the buyer beware!*

Unwrapping your new gizmo is part of the fun of buying it, but don't misplace anything in the process. Some adapter plugs are small enough to get tossed along with paper and cellophane. It is possible for any electronic equipment to pass stringent quality assurance tests and then fail shortly after deployment in a phenomenon known as *crib death*. As a general rule, we don't throw anything away until the equipment has been up and running for at least a week. Should it become necessary to return parts, the manufacturer will expect to get them back in their original box. The instruction manual will contain a list of the parts you should have. It is a good idea to check the box's actual contents against that list. If something is missing, the manual will also contain a number to call for questions. Chances are that your manufacturer will be more agreeable to replace pieces or the entire box before you have used, or attempted to use, the contents for several days.

If you've purchased several items at the same time, then put the parts for each in separate piles on the floor. One power transformer or patch cord looks a lot like another when they are set down together. Should you accidentally switch parts, one of your new modules might not work afterward.

Scrutinize and Familiarize Take a moment to take a look at the parts for any dents or scratches. The first few pages of the instruction manual will show you simplified diagrams of the front and rear of your new Wi-Fi access point. The layout of access points and wireless routers has become fairly standard, with connecting cords draped outward from the rear, and the indicator lights on the front. Sometimes the labels on connector ports are not easy to read, especially if they have been stamped into metal, so invest the moments necessary to hold your router up to the light. Compare it against the diagrams so you can see what plugs into what. The RJ-45 LAN sockets, for example, are apparently identical, but one may be wired internally as an uplink port for attaching a wired hub and will not work if you plug a PC directly into it. Likewise, another will be dedicated for connecting to your cable/DSL modem and will operate only at 10 Mbps.

When the router is in place with wires attached, it will be difficult to turn it all around to read these labels from the rear. You may want to tape a label of your own on the top, with large print and arrows, so you can tell at a glance where the cables should properly be plugged. At some point in the future, you may wish to pull the router out to set up a temporary WLAN somewhere else, for example. Labels on the cables and the router make reinstalling it in your home afterward a simple task.

The Warranty The first day is also the best day to note the model and serial numbers, which you may have to do in any case in order to fill out the warranty and registration card. Include the date of purchase on your copy. This is the first step toward documenting your new system. We've never been refused service or replacement for lack of a warranty card, but that's no guarantee that your manufacturer might not ask for proof that you are the original owner. Vendors may save this information so that if you forget when you bought your router, they can tell you. If you are more comfortable with a keyboard than a pen and don't want to spend a postage stamp, most vendors will let you register your product online. In fact, they prefer it, since the information you provide will not have to be transcribed by a third party.

Modern manufacturers, particularly high-tech sellers, know the value of information about their customers. Data gathered from a sale represents a golden opportunity to promote an ongoing relationship and future sales, as well as guidance for new product development. Theoretically, however, you don't have to fill out the questionnaire that comes with the warranty. You don't have to tell anyone how much money you make, your age, or anything else to get a broken part fixed. You will have to provide a postal or email address, but as mentioned earlier in this chapter, you can have multiple email addresses and dedicate one of them for tasks like this. Vendors say they need an address in order to inform you about patches or upgrades, but don't be surprised if you get spam along with the notices. Upgrade information is something you can find on your own by periodically consulting the vendor's web page. Remember that whatever information you volunteer can be resold, copied, and echoed across the Web to many strangers for years to come.

Host Hardware Minimums Most manufacturers will put minimum hardware and software requirements on the box so that consumers can check in advance of their purchase. As with all peripherals, your PC's hardware must be powerful enough to accommodate the new addition. Hardware requirements are rarely a concern these days as most modern PCs come with more than enough firepower. But if yours is an older machine, if it is heavily loaded with other peripherals, or if you run many programs on the desktop at once, you should check the box first. Also, you should know the type and version of your PC's operating system and the version of your web browser as well. Your browser is the tool you will use to configure your new router.

Here are some examples of minimum hardware requirements:

- Processor: 200 MHz or faster Pentium
- 64MB memory
- CD-ROM drive
- Specified operating system: Windows 98, ME, 2000, and so on. If you have a Mac, check to see that it is supported.
- If it's a USB interface, you should have an open USB port.
- Tip: If it's a *Peripheral Component Interface* (PCI) card, older PCs won't accommodate it if they only use ISA (Industry Standard Architecture) sockets.

It's wise to read the box-stated minimums, because they will vary from part to part. If your PC skirts the minimum requirements now, then your wireless experience (and your user experience generally) will be much improved with some PC hardware upgrades. The cost of computer memory, for example, has dropped to a level thought impossible a few years ago: less than a dime per megabyte for standard types. If your disk activity light shows that your hard drive is active continuously as you work, your PC is likely swapping program segments from main memory into and out of the hard drive. Onboard memory is much faster. Online retailers will ship it to your door in days.

It is possible to load software from a CD-ROM if you do not have one built in. If your PC is already on a home network, you can share another PC's CD-ROM and read from it across the network as though it were a shared hard drive. (We'll detail the steps to sharing a drive later in this chapter.) Since in this case you are loading network software, it would be safer to copy the contents of a network-shared CD onto your own hard drive, if you have the space, and execute the installation from there.

Read the Instructions It is sometimes jokingly said that "real engineers don't read instructions" in the same way that "real men don't ask for directions." The truth, however, is that professionals always open the manual before they open their tool case. Equipment vendors realize that most customers are reluctant to read the fine print with a new gadget at hand, so they often provide a quick-start guide that unfolds like a poster or road map. They do so because new equipment always comes with a few cautions, which, if ignored, can result in damage, frustration, or dissatisfaction with the product. This is particularly true for wireless, whose technology is continually being refined. Even if familiar equipment is being reinstalled, the new host PC may be running an updated or entirely different operating system.

Here is one point from the manual you must deal with before you start: Some Wi-Fi equipment must be physically attached to a PC before the driving software is loaded from a CD-ROM. Others demand that software be in place before the PC is allowed to discover the new equipment. Newer PC operating systems will discover your Wi-Fi router and configure themselves without drivers from a vendor's CD-ROM. But you can't know specific software requirements just by looking at the router. Looking over the first few pages of the install guide may save you hours of avoidable deinstallation and reinstallation. It should be your first, not your last resort.

For simplicity, we will break down the wireless router installation into five steps:

1. Copy the configuration settings from your existing equipment or query your ISP.
2. Physically connect the router to your broadband network.
3. Physically connect your (wired) workstations to the router.
4. Configure the wired and wireless workstations to use the router as well as file and printer sharing.
5. Configure the router for your ISP and security needs.

We will assume that your cable/DSL modem is already installed and working.

ISP Configuration Settings

For your PC to connect to the Internet through your cable/DSL modem, it must have an Ethernet network interface of some kind. It may be a built-in network socket on a laptop or a *network interface card* (NIC) installed on a desktop. That interface provides the data pipeline to your cable/DSL modem. Your PC will also use special software that initiates and controls that Internet connection. You may remember that your ISP sent you the software, probably on a CD-ROM, which you loaded into the PC. Afterward, you had to supply your PC with some special numbers or *configuration parameters* that were unique to the ISP and the PC.

Start with a Working Connection If your PC is not talking to the Internet, this would be an excellent time to call your ISP's technical support line to get it working. Unless you plan to use your wireless net in isolated, peer-to-peer ad hoc mode, an operable Internet connection is a necessary foundation for the expansion of your network. A dead connection will not come to life simply by adding more equipment on your end.

Copy What Works Your Wi-Fi router has the necessary connection software built in already. If you give it the same parameters that you entered into your PC to get it working, then the router can imitate the PC perfectly. You will probably be able to copy the parameters you need from the working PC. This information can be found from two other sources. Most ISPs have online tutorials, frequently asked questions, and walk-throughs specifically tailored for new customers. You may want to print this material out before you disconnect your hardwired PC from the modem. As a last resort, you can always call the ISP technical support line. They probably have technicians who specialize in dealing with confused first-timers. It is not an unusual scenario.

Take a moment to observe the indicator lights on your cable/DSL modem when it is operating normally. Make a sketch that you can refer to later on for comparison when you've replaced the PC with your router. That makes it easier to tell when the system is working as it should.

Removing Old Point-to-Point Protocol over Ethernet (PPPoE) Software Some equipment vendors suggest that you remove the old connection software, that is, EnterNet or WinPoet, before you use your PC as part of a local, router-based network. Our tests showed no problems after we left it in place, as long as it was not configured to start automatically at boot time. But if you like your system to be clean and simple, here's how to deinstall it:

1. Click Start, then Settings, and then Control Panel.
2. In the Control Panel window, find Add-Remove Software and click it.
3. A list of your installed software will be displayed in the window that opens. Scroll down until you find your ISP-provided software and highlight the name by clicking it once.
4. Click the Add/Remove button. You may have to reboot afterwards.

You will need the capability of running your PC as a standalone with the cable/DSL modem in order to troubleshoot any network problems you encounter in the future. ISP technicians will not investigate a defective Internet connection that relies on equipment they did not provide. They will ask you to remove your router or access point first, and then reconnect your cable/DSL modem directly to a single PC before proceeding (see Chapter 6's, "A Case History"). When you are finished removing the software, put the installation disks for it, along with your configuration notes, in a safe place.

What's My IP Address? Every node on the Internet is identified with a unique number, including your PC and eventually your new router. (See "Internet Addressing Basics" earlier in this chapter.) If you've ever used a telephone modem for a dial-up connection, you know that your IP address was automatically assigned to you at logon only for the duration of your session. Someone else recovered it for reuse afterward. Put another way, the address was *dynamically assigned*. Most ISPs have continued this approach after accepting broadband customers. Even though you don't hear the funny noises a telephone modem makes when dialing, your PC and your router log into your ISP over your cable/DSL line, and most will be assigned a temporary IP address as a result. In your research, then, you must determine if you have a dynamic or static IP address. Your ISP can tell you that. If it is static, it never changes (and you probably pay more for that). If it is static, you can find out what it is by surfing your hardwired PC to www.whatismyip.com.

An IP address is not enough by itself. A *subnet mask* is assigned by your ISP to determine how that IP address will be applied. It will be in the same format as the IP address, perhaps 255.255.255.0. The router will also need the IP address of the *default gateway* on the ISP's end of the broadband line. These parameters will not be the same as the ones for your telephone dial-up connection.

Before you disconnect your PC, Windows users can see a display of these parameters by running the program WinIPcfg (see Chapter 6). If you happen to be running Windows NT or 2000, you can find the same information by opening up a command window (in the Run dialog box enter the command CMD) and then enter the name of the text-based equivalent program: *ipconfig /all*.

Other ISP Connection Parameters No universal method exists for connecting a PC or router to a service provider, although the majority of providers use an extension of the telephone dial-up modem method, known as the *Point-to-Point Protocol* (PPP). On broadband connections, this is known as *Point-to-Point Protocol over Ethernet* (PPPoE). If this is the case with your provider, you must know your login and password. These will be the same as for your dial-up connection, usually in the format yourname@provider.net. Some other items you must determine are as follows:

- Do you have a *hostname* assigned by your ISP, and if so, what is it? The WinIPcfg program can reveal that too.
- Does your provider assign a *service name*? Not all do.
- Does your connection expect a particular *MAC address*? This is a unique identifier consisting of a manufacturer's ID number plus the serial number of the network device. Cable/DSL modems can be set by your ISP to accept data from one particular device. If this is the case with yours, it will stop working when you attach your router in place of the PC NIC. Fortunately, routers can copy (or *clone*) the MAC address of your PC's NIC. You can find that number using the WinIPcfg tool also. It will appear as six hyphenated pairs of hexadecimal numbers, such as 00-11-22-33-44-55.
- The IP address of the *domain name server* (DNS). This facility acts like a telephone directory to translate the names of resources into their IP addresses.
- You may need to know the full and exact IP address of the ISP's email server or home page.

Insert the Router into Your Network

You've successfully extracted the information you need from your working PC. Now we want to replace that PC with your new wireless router, so it can connect to the Internet as the PC did. In addition, the router will distribute that Internet connection among multiple PCs in your home. To replace the PC with the router, follow these steps:

1. First, turn off the power to all of your equipment.
2. Take the patch cable that connected the cable/DSL modem to your PC, disconnect it from the PC, and reconnect it into the proper port on the router instead. This router socket may be labeled WAN, for *wide area network*, or DSL.
3. You will need another LAN cable to connect the PC you were using to the router. One end goes into the NIC socket on the PC's LAN interface, as before. The other end will go into any open RJ-45 socket on the

router, as long as it is not labeled uplink or marked with the rollover symbol, which resembles a horizontally stretched *X*. An uplink port is used to connect a hardwired hub, so its wiring is internally reversed.

Connect Other Hardwired PCs

In the same way, any other hardwired PCs will be plugged into open LAN ports on the router. One port is as good as another. They may be labeled 1, 2, 3, and so on, but none has precedence or more speed.

Older Ethernet NICs run at 10 Mbps. They will do so over older CAT3 or the newer CAT5 LAN cables. The router autosenses the transmission speed and adjusts its port accordingly. If your PC has a newer, fast Ethernet NIC that is capable of 100 Mbps, then the patch cable you use to connect it to the router must use CAT5 wiring, which is fast Ethernet cabling. Use a connecting cable of an appropriate length. If a long patch cord is coiled up on itself for a short connection, then the speed of the connection may drop. Shorter is better. If you plan to use one of your PCs as a local server of some kind, it should be connected using a fast, wired interface card.

Your PC LAN connection may not necessarily be through a network interface circuit card in a desktop PC. It might be through a built-in network port on a laptop or through a USB adapter. The form factor is unimportant. To your router, they will all look the same. Continue the hookup process until all your wired PCs are plugged in.

Power Up If you haven't already done so, this is an ideal time to deploy a surge suppressor or UPS, so that your new equipment will be protected from spikes and sags in the incoming power (see Chapter 5's "Power Corrupts"). The outlets farthest from the power card have the best protection.

Insert the transformers or AC cables to power up your network in this order:

- Cable/DSL modem
- Wi-Fi router
- The workstations

First Troubleshooting Take a close look at the indicator lights on your equipment to see how you've done so far. The AC or power lights, called *light-emitting diodes* (LEDs), should be lit, of course. After the cable/DSL modem completes its power-on diagnostics, the light indicating a connection from your cable/DSL modem to the Internet should be on as you noted earlier, because you haven't unplugged or changed anything in that direction.

When two devices are properly connected across a network, each will have a simple indicator LED, usually green in color, that lights up when a signal is received from the other. Other LEDs may exist as well to indicate data activity or errors, or the green light may blink to indicate network activity, but a connection or sync light should be on at both ends of the connection from the modem to the router. Each of the router ports will have one of these link lights, as will the NICs on each of the wired workstations. The instruction manual will show you where to look. If a connected pair is conspicuous by its darkness, try swapping ports, PCs, or cables methodically until you determine the weakest link. You can determine which port is connected to which PC in this manner by momentarily unplugging the patch cord between them in a process known as *blinking*. After identifying ports and cables, label them if you haven't already.

Configure the Workstations

The automated assistant software provided on the CD-ROM that came with your equipment can configure your workstation for you if you are willing to live with the defaults you've been given. That's acceptable in order to get it working in the beginning, but you should not do so and walk away. To ensure the security of your system, you will have to adjust some parameters on your own after the initial setup.

As with wired PCs, your Wi-Fi network interfaces require standard IP information to participate in your local IP network. (They also require a few extra parameters in order to manage their wireless link.) If your network uses equipment from different manufacturers, if you assign your own IP addresses, or if you are accommodating a nonstandard setup for any reason, you will have to enter some networking information yourself. The manner in which this is done will vary somewhat from one operating system to another. The following sections outline some examples.

Workstation Addressing for Windows 2000 or ME Right-click the My Network Places icon. Select Properties from the drop-down menu (see Figure 4-11).

You have two ways to proceed: either assign a permanent address yourself or have your new router assign a temporary address. The DHCP capability in your new router is identical to the process used by your original

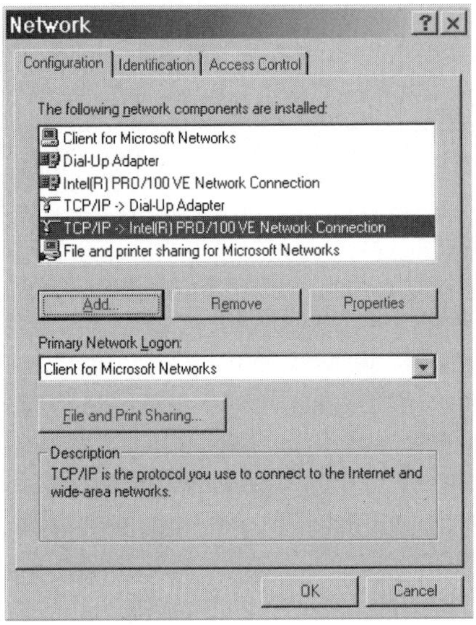

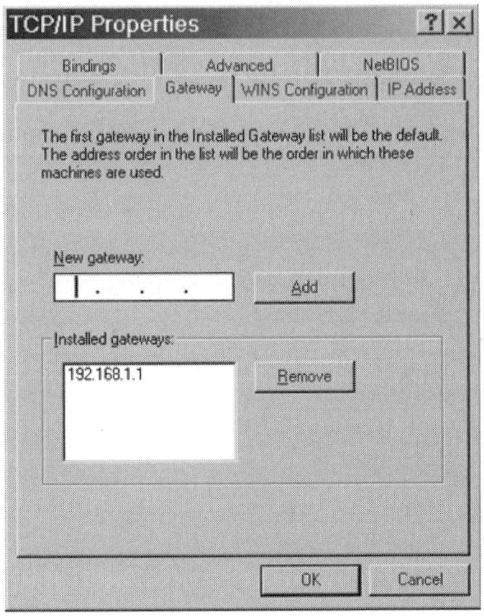

Figure 4-11
IP configuration screens

hardwired Internet PC to download a temporary IP address at login time. The easiest method (and the default) is to let each of your PCs ask the Wi-Fi router for an IP address when they boot up. But one disadvantage is that you might not get the same address every time. The local IP addresses are passed out on a first-come, first-served basis. For most applications, this is not a problem. However, if you want to designate one of your PCs as a gaming, web server, or video conferencing PC, it must have a predictable and static IP address in order to be reliably exempted from the router's built-in firewall blocking.

Actually, there's no reason why you can't act as your own ISP and assign a permanent IP address to all your workstations. It's easier to keep track of them and troubleshoot problems that way and harder for an outsider to guess what those addresses might be. The addresses you choose must be within a range acceptable to your local Wi-Fi router (see Figure 4-12) and your instruction manual will tell you what those are. Of course, every PC on your home network must have a unique network address. But all of them will share the same subnet mask, DNS server address, and gateway address.

In this case, your local network's gateway to the Internet will be through your Wi-Fi router. This gateway will serve as your computers' DHCP or IP address server. Your new router was assigned a default IP address at the factory. In theory, you don't have to specify your new router as the DHCP

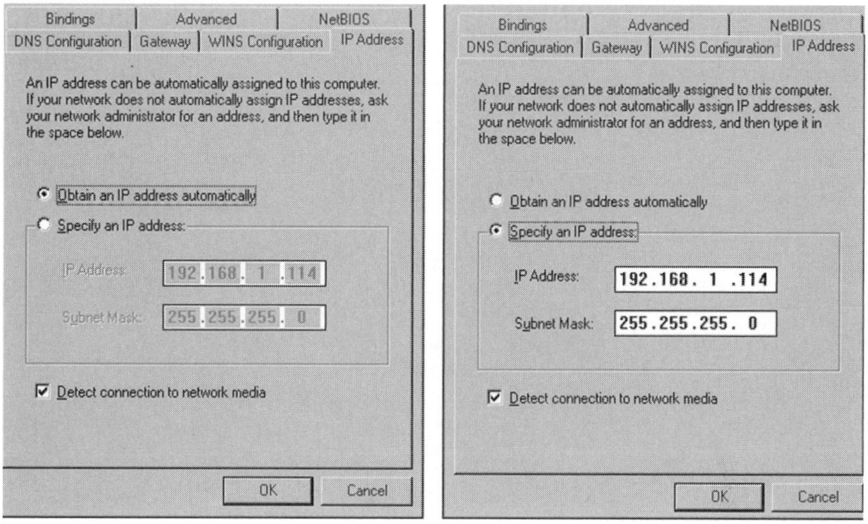

Figure 4-12 Assigning a permanent IP address to your workstations

source, but doing so will prevent the PC's network software from looking for a source at 10-minute intervals.

Manual IP Configuration in Windows NT The entry screens appear differently for Windows NT workstations or servers, but the same IP parameters must be entered for each Wi-Fi interface. To do so, follow these steps (see Figure 4-13):

1. Start by clicking the Control Panel icon and then Networking.
2. Select Protocols, then TCP/IP, and then Properties.

You can enter a permanent IP address and subnet mask if you want, or say that you want the PC to pull an IP address from your router/gateway. (The values you see illustrated are for a permanently assigned IP address and are for examples only. You will have to look to your manual to find appropriate values.) When you have completed, your PC will ask that you reboot before the changes take effect.

To configure the IP parameters for a Mac, perform the following steps:

1. Select the Apple menu.
2. When it drops down, select Control Panels.
3. Select TCP/IP. When this window opens . . .
4. At the Connect VIA line, choose either Ethernet Built In or Ethernet, as appropriate for the network adapter type in your particular machine.

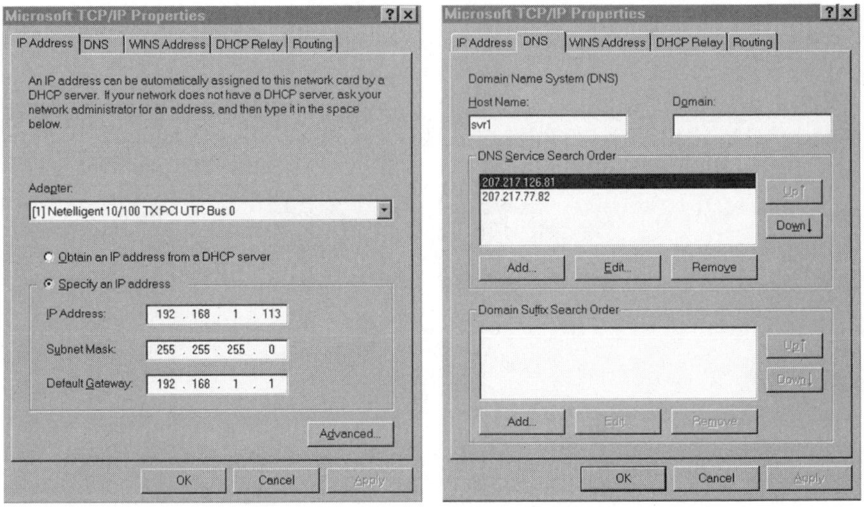

Figure 4-13
The configurations screen display on an NT machine

5. At the Configure line, select Using DHCP server so the Mac will get its IP addresses from your Wi-Fi router at boot time using DHCP. If you want to assign a permanent local IP address, you will have to enter those parameters also, as with the previous Windows machines.
6. It will ask if you want to save the configuration. Click Save.

Workstation Networking Parameters

When you had only one PC attached to the Internet, matters pertaining to local networks were irrelevant. If all the PCs you add to that connection will communicate only to Internet resources, you won't have to set them up for local networking or file sharing. But local resource sharing is a major benefit of having a router. It is even more useful with mobile Wi-Fi workstations. It gives you the ability to copy files at high speed between your PCs, and to share a single printer among several PCs. When all your home PCs work together, they form a network that is greater than the sum of its parts.

As you determine what is shared and to what degree, you take on the duties of a *system administrator*. You do not necessarily have to sit at a particular PC in order to do this work. Later versions of Windows enable you to configure networked PCs remotely if you first allow remote administration (through the Passwords Properties tab in the Control Panel). But you probably won't need this feature in a home environment where you can easily walk from one computer to another. It is an increased security risk also, since someone connecting to your network from the outside might be able to use it to gain further access.

File Sharing File sharing is one of home networking's killer applications. By using it, any PC on your network can use storage on another's hard drive (or CD-ROM) as though it were its own. Setting it up requires three components. To see if the first of these ingredients, the Client Service, is presently active on a PC,

1. Right-click the Network Neighborhood icon on the desktop.
2. Select Properties from the menu.

You should be able to see Client for Microsoft Networks in the white box (see Figure 4-11, left panel). The initial Windows setup adds the necessary software as a default, but if it did not or if the feature has been disabled, you'll have to put it back:

1. Click Add in the network Properties window you just opened.
2. Pick Client from the list of choices in the popup window that appears.
3. Click Add. You are presented with a list of different software vendors.
4. Click Microsoft. This displays a listing of Microsoft's services (see Figure 4-14).
5. Pick Client for Microsoft Networks from the list and click OK.

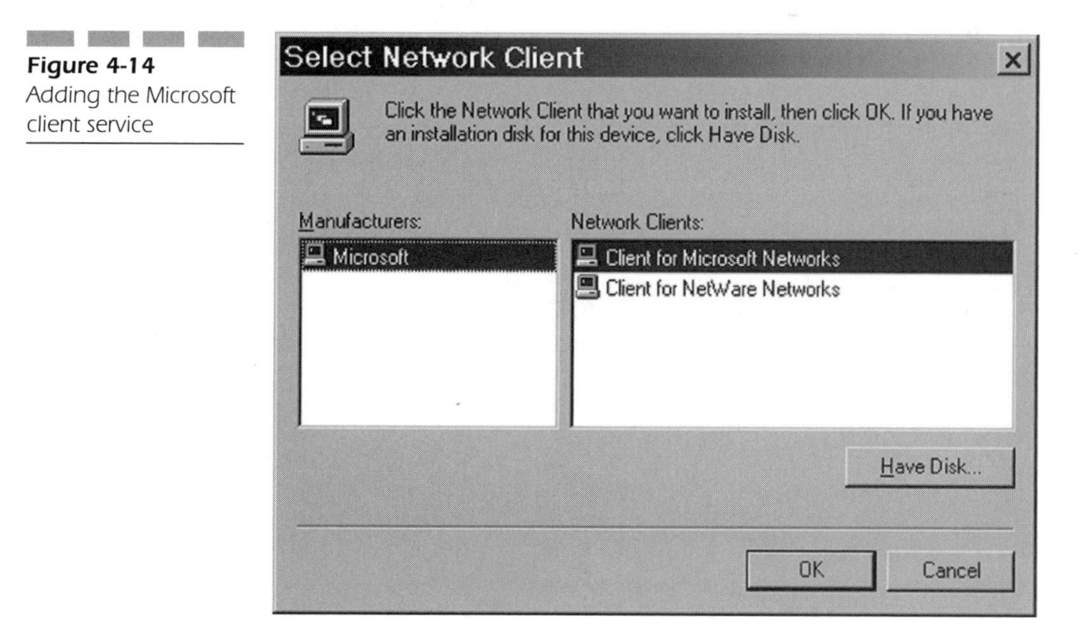

Figure 4-14
Adding the Microsoft client service

You may have to insert your original Windows installation CD so files can be copied from it onto your PC. The second component is the installation of the file sharing service. From the Network Neighborhood's Properties window,

1. Click the Add button, and select Service as the type of network component you want to install. Click again on the Add button.
2. The service screen will give you the choices. The Network Service you need is *File and Print Sharing for Microsoft Networks*. (See Figure 4-15).
3. You are presented with two options: Allow access to files, and allow others to print. Click the boxes for both. When they are enabled, they will display a check mark.
4. Click OK to close the Sharing Options window.

Deciding Remote Access Capabilities Now that sharing is possible on your PC, it's time to determine the manner and degree. Windows gives you two basic sharing methods. You can choose to share resources with specific, predetermined, named PCs in a process called *user-level access control*. In this way, one computer on the net will carry a list of PCs or domains that will predetermine users' access rights. The feature is designed for large networks and is not necessary for home use. A simpler method is to choose *share-level access control*. With it, any PC on your home net with the proper password can get to a shared resource. To set up this last file sharing component,

1. Click the Access Control tab near the top of the Network window.
2. Click the box beside Share-level access control (see Figure 4-16).

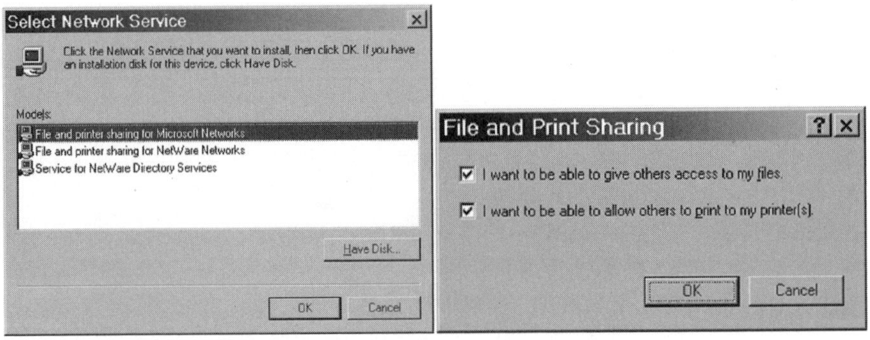

Figure 4-15
File and print sharing is enabled.

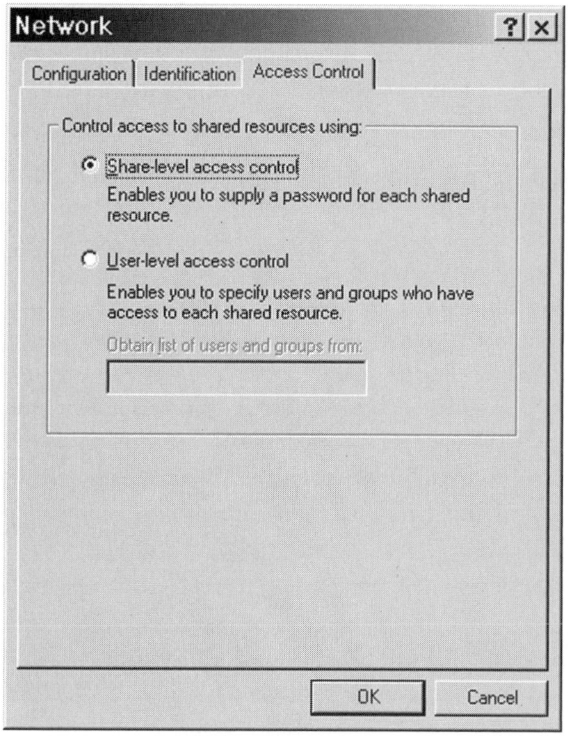

Figure 4-16
Selecting share-level access

Your PC and Workgroup Name Every PC on your network is identified by a unique IP address, but it's easier to refer to them by an individual name that you give them yourself. In addition, each station will belong to a *workgroup*. If you have many workstations, you can divide them into several workgroups, but this is unlikely in a home environment. Unless you own a PC farm, one workgroup will suffice. You probably will want to change that workgroup name from the Windows default (the word *Workgroup*) to something unique for the sake of originality, if not security. Pick something that an outsider will have difficulty guessing. Here is the naming routine for Windows 98/ME:

1. Right-click the Network Neighborhood icon on the desktop.
2. Select Properties from the menu.
3. Click the Identification tab. You are presented with three boxes to fill in.
4. In the first box, enter your computer's name. It must be unique.

5. Your workgroup name goes into the second box. Unlike the unique computer name, all your computers must have the same workgroup name, exact in every letter (see Figure 4-17).
6. A comment is not necessary, but as your own system administrator you should give yourself every advantage by adding a little explanation here. PCs tend to look alike when remotely viewed as icons.

Sharing a Windows Folder Now that your computer has file sharing installed and a name to share by, we can get down to the business of determining what specific resources to make available to your local network. You do this task on a PC-by-PC basis. You can share one particular folder on a PC or, if you want, the contents of an entire hard drive. When you share a folder, all the subfolders in it will also be shared. If you share a drive, the contents of the entire drive will be shared. Naturally, the less you share, the more secure your network will be. And if you give wide-open access to

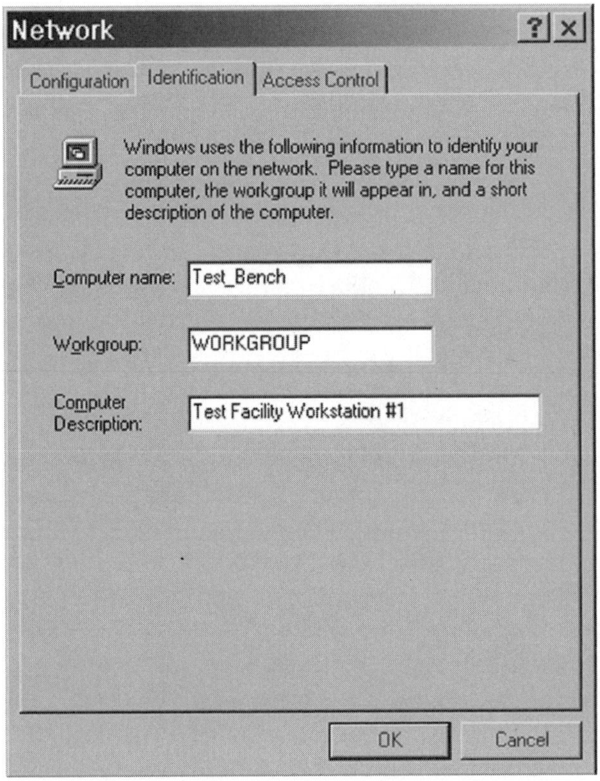

Figure 4-17
After installing the file sharing software, pick names for your PCs and their common workgroup.

a hard drive, a remote user can accidentally delete an enormous amount of information with a single mouse click. Denying access to some files on some PCs is standard practice for corporate businesses and should be for your home office. As a specific example, your children have no business in your financial records and your PC probably contains other documents they won't need to change or even see.

When you do share an item, you may specify a password for access to it, and it's a very good idea to do so. After their first access to the resource, your other PCs will remember the password if you want so that you won't have to. To share a folder, follow these steps:

1. Right-click the icon for the folder you want to share.
2. Left-click Properties.
3. Select Sharing.
4. Unselect the default button Not Shared by clicking instead on Shared As.
5. The name by which this offering will be known to the other PCs on the network defaults to the name of the folder or hard drive. You can call it whatever you like (see Figure 4-18). You may want to include the name of the PC to make it clear to others where this particular resource is located. The comment field is good for this purpose also.
6. Select the type of access you are willing to grant. *Read-only* limits visitors so that they cannot change the folder's contents. *Full* enables them to do whatever you may do from that PC's keyboard, including deleting files.
7. You can selectively control network visitors' rights by giving particular users either a read-only or change-allowed password. Those without a password won't get in.
8. When completed, click the Apply button. The resource becomes available at once. Its icon will appear with a hand underneath, confirming that it is offered as a shared resource.

Accessing a Shared Folder One important point to remember about remote file access is that personal firewall software will intentionally block it unless you specifically allow otherwise. Trying to establish a connection with another PC (or with your router) the first time can be a frustrating experience if you don't reconfigure the firewalls on both ends. If you can ping a device but cannot connect otherwise, disable your PC firewall temporarily to see if it is the cause.

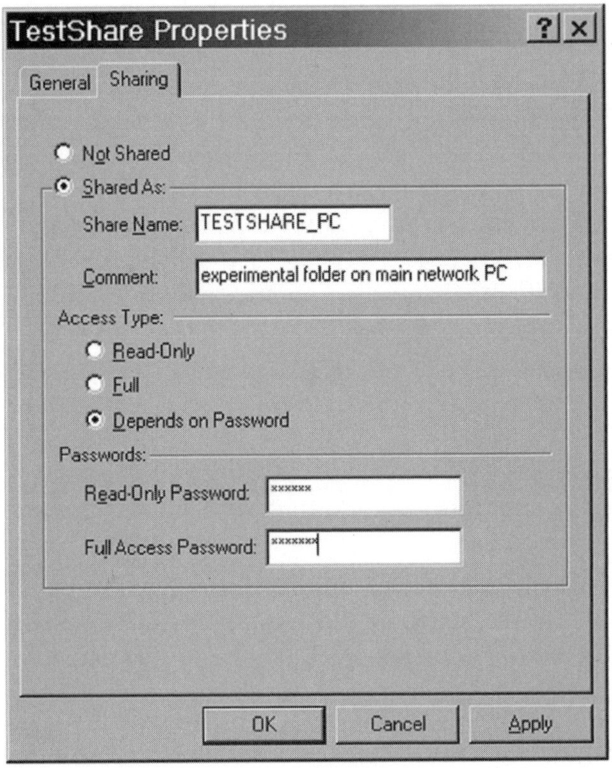

Figure 4-18 Sharing a folder with other workstations on your network

Getting information from a shared folder can be done in two ways. You can set up a shared resource on a remote PC as a virtual hard drive for your own and even create a shortcut for it on your desktop:

1. Right-click either the desktop My Computer icon or the My Network Places icon.
2. Click Map Network Drive.
3. On the pop-up window that appears, you will have two boxes to fill in. The top one is the virtual drive letter you will assign to the resource. The drop-down menu will show you which letters are available for use.
4. The second line is the name of the resource. This consists of two backslashes, the network name of the PC on which the folder resides, and the share name of the folder itself (see Figure 4-19).
5. The checkbox gives you the option of automatically reestablishing this connection every time your PC boots up. This is useful for items that

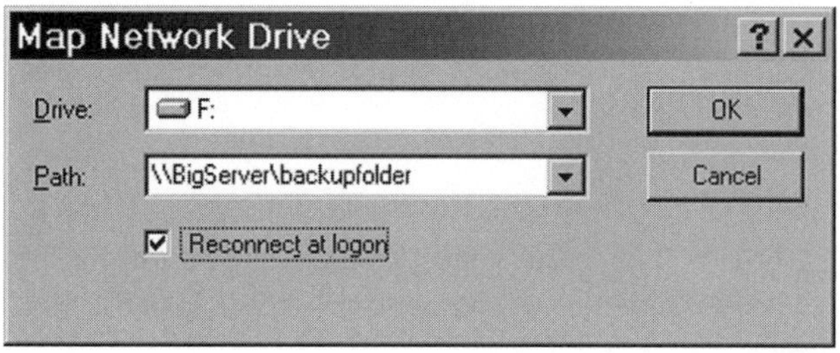

Map Network Drive

Drive: F:

Path: \\BigServer\backupfolder

Reconnect at logon

OK

Cancel

Figure 4-19
Mapping a shared folder as a virtual disk drive

you will access frequently, such as a backup directory that will be used every night.

Later versions of Windows provide an even simpler, GUI-based method of remote access:

1. Double-click Network Neighborhood with your left mouse button. This opens a window that displays icons representing all the PCs on your new LAN. If they've just been turned on, it may take a few moments for them to show up. Wi-Fi PCs look no different from the others.
2. Double-click the icon for the computer you want to access. Another pop-up window will show a list of the resources you have shared on that PC.
3. Double-click the desired folder or drive.
4. If this is your first access, then you will be prompted for the password. Enter that password, and you are in.

Sharing a Windows Printer Printer sharing is a popular and efficient application of home networking. Wi-Fi manufacturers sell wireless peripherals for just this purpose, and sometimes shared printer ports are built into wireless routers as a value-added feature. It is possible for any network-attached PC to share access to its hardwired printer with any or all of the other PCs on that home network, including the wireless ones. They will be able to use the printer as though they were exclusively wired to it themselves. Naturally, the computer to which the printer is actually attached must be booted and running. We will assume that you've installed File and Printer Sharing as described earlier. To share a printer, perform these steps:

1. Click Start in the lower-left corner of your desktop.
2. Select Settings.

3. Click Printers.
4. Select the printer you want to share by highlighting its name.
5. Go up to the File menu to click Sharing.
6. Click Shared As.
7. Here, as with directories, you can specify a name by which this printer will be known to other network users (see Figure 4-20).
8. You can also demand that a password be supplied to use your printer. Printers store no information, however, so their threat to your network's security is minimal.
9. Click Apply. The printer's icon will now show a hand underneath, meaning it has been successfully offered as a shared resource.

Using a Network-Attached Printer Remote printing from a Wi-Fi laptop is very handy. You must add the printer to your PC's configuration only

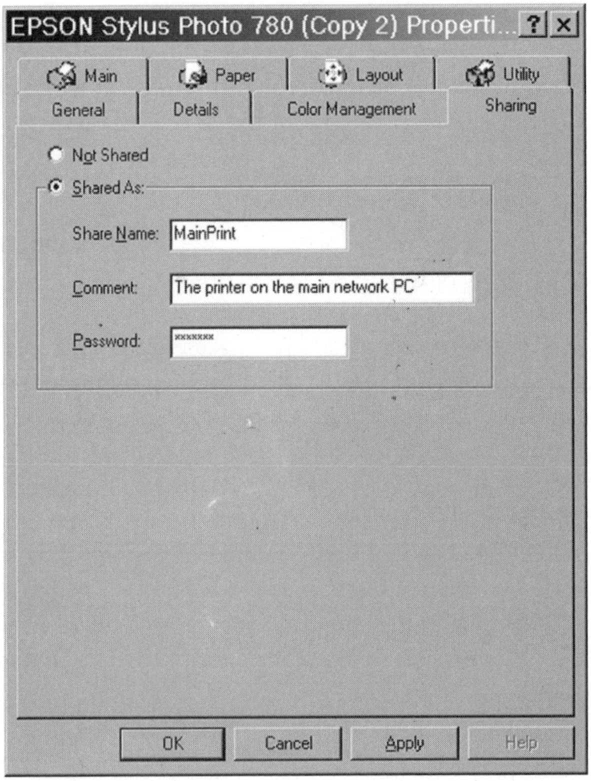

Figure 4-20
One network and one printer for all

once, which is easier than repeatedly disconnecting a printer cable from a PC and reattaching it to a laptop for a few moments' use. Most modern printers are complex devices that require specific software to control and format their output. They also continuously report their status back to the PC that is using the printer. You will have to install drivers for your particular printer on every PC that will virtually connect to it. To access a printer from another computer,

1. Open the Control Panel Icon by clicking it.
2. Open the Printers icon.
3. Double-click the Add a Printer wizard.
4. Choose the Network Printer option and click Next.
5. If you don't remember the share name of the printer, the wizard can display a list of all known printers. Choose the one you want and click Next.
6. If the printer software is already on the PC, the wizard will install it from there. If not, insert the CD with the printer's driver software so that it can be copied over.
7. You will be given the option to print a test page. You can do this anytime, but it never hurts to test.

Loading and Configuring Workstation VPN Support

VPN is a method of thoroughly encrypting data so that it can be sent over the public Internet (and Wi-Fi airwaves) with confidence that it will not be compromised if intercepted en route. Remote workers who need to access a corporate intranet most commonly use VPN. (For a general explanation of how VPN works, see Chapter 7, "Wi-Fi for Fun and Profit.") If you are installing VPN software on a laptop that you intend to use at home, you may want to visit your company's Internal Telecommunications department and ask them to do it for you. You will have to visit them in any case, because they will have to supply a company-standard version of the software (there are many kinds) along with logins, passwords, and IP addresses.

That being said, to install VPN on your PC yourself, first ping the corporate network access server to verify that you can reach it. (That IP address is one of the items you will need from Corporate.) Then add the VPN software to your PC from your Windows install CD. Here's the routine for Windows ME:

1. Start at Start. Click Settings, then Control Panel, and open the Add-Remove Software Icon.
2. Click the Windows Setup tab. Scroll down to the Communications icon. Double-click it.
3. A second scroll window appears. Find and check Virtual Private Networking.
4. Put the CD in the drive and click OK twice. The software will be copied and you will be prompted to reboot your PC.
5. You may want to check the Microsoft web pages to see if any new VPN patches are available.
6. Again, open up the Control Panel. This time, select Dial-Up Networking.
7. Click Make New Connection. Enter a name for your corporate office.
8. Choose the Microsoft VPN Adapter as your device.
9. For the IP address of your VPN server, enter the IP address for your office server, the one you pinged earlier.

To use your new VPN, simply double-click the VPN connection icon in Dial-Up Networking. You will be prompted for the user ID and password given to you by the software provider (see Figure 4-21).

Wi-Fi in Windows XP

The latest rendition of the Windows operating system was built with wireless networks very much in mind and contains several features that

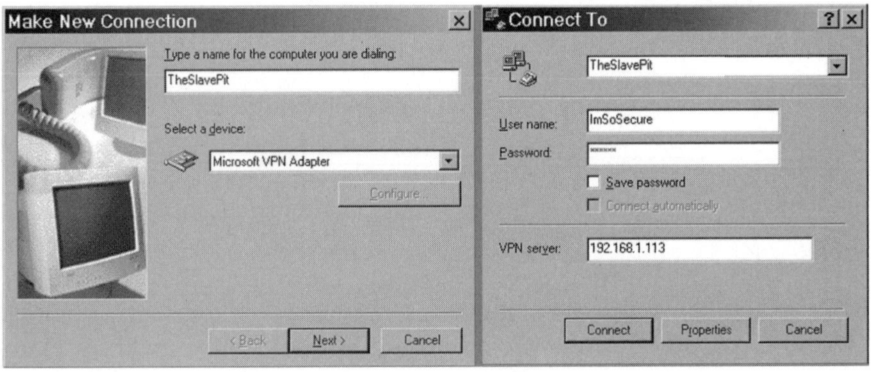

Figure 4-21
Configuring and using a VPN connection

make working with them simpler than in the past. One of those, as mentioned earlier, is Internet Connection Sharing (ICS). If you plan to use a Wi-Fi router, you will not need this feature. If you plan to use a wireless access point to connect several Wi-Fi PCs to the Internet, this software can make a PC perform like a router. Microsoft cautions that you shouldn't use this feature in an existing network that already employs Windows domain controllers, gateways, DHCP or DNS servers, or with systems configured to use static IP addresses.

Connection Sharing for XP ICS serves the same function as the software router programs mentioned earlier. With ICS, you connect one of your computers to the Internet and then through it, you may share that Internet service with other computers on your SOHO network. The Network Setup Wizard in Windows XP Professional will automatically provide all the network settings you need for this task.

To use ICS in place of a hardware router, you must first enable it on a PC that has a connection to your cable/DSL modem. As with hardware routers, you will have to configure your client workstations to ask the ICS PC for an IP address at boot time. You must also configure the Internet options for ICS.

To enable ICS Discovery and Control with Windows 98, Windows 98 SE, and Windows ME, you will need to run the Network Setup Wizard from the install CD. Before you do, check that your version of Explorer is 5.0 or higher. You will have to be logged onto the Internet-connected PC in an owner account:

1. Start by clicking Start.
2. Then click Control Panel and double-click Network Connections.
3. Select the LAN, VPN, or PPPoE connection you are going to share.
4. Under Network Tasks, click Change settings of this connection.
5. Click the Advanced tab. Check Allow other network users to connect through this computer's Internet connection.
6. You can allow network users to turn sharing on and off by checking Allow other network users to control or disable the shared Internet connection.
7. In Home networking connection, under Internet Connection Sharing, select any NIC that connects your Internet-connected PC to the other PCs on your home network. (The Home networking connection is only present when more than one NIC is installed on this computer.)

Here's how to configure ICS options on your client computers. First, check the configuration of your Windows browser and Internet connection:

1. Open Explorer. (Click Start, click All Programs, and then select Internet Explorer.)
2. Click the Tools menu and click Internet Options.
3. On the Connections tab, click Never dial a connection and then click LAN Settings.
4. Clear the Automatically detect settings and Use automatic configuration script boxes. In Proxy Server, uncheck the Use a proxy server check box.

XP Automatic Roaming XP has built-in facilities to accommodate the mobility of a wireless user. It can detect Wi-Fi networks, force a reauthentication, and even detect and adjust for a new IP subnet. Users can prearrange dynamic or static address configurations and the appropriate one will be chosen automatically as the user wanders. Internet Explorer's proxy settings and network firewalls can also update themselves as their base network changes.

Windows XP will configure a Wi-Fi NIC to scan the airwaves for available networks and then associate with one when it is discovered. If more than one exists, a user can predetermine the order of preference or skip those that aren't preferred at all. If none are found, XP either defaults to ad hoc mode or else disables the Wi-Fi interface as determined by the user. If you know that a wireless router or access point is in range but it does not show up in the list of visible networks, you may still specify it as the preferred connection:

1. Again, start with Start.
2. Open the Control Panel.
3. Open Network and Internet Connections, and then select Network Connections.
4. Right-click Local Area Connection.
5. Click Properties.
6. Verify that the Show Icon in Taskbar Notification Area When Connected box has been checked. This nifty feature lets you see the status of your Wi-Fi link by putting your mouse pointer on it.
7. Click the Wireless Networks Configuration tab. You are shown a list of known wireless networks. Click Add.

8. Type the name of the access point you want to add into the Network Name or SSID box. Specify your WEP settings if you are using any, including your key.
9. Click OK.

Wi-Fi Workstation Configuration

The manufacturer of your Wi-Fi interface cards will provide instructions and software that load automatically into your wireless workstations. These programs will autoconfigure the Wi-Fi interface using default values. If you want to make your network your own and enhance its security, you will have to manually enter unique IP addresses plus the other parameters that you entered for the wired PCs. Their radio links are independent of their IP addressing requirements. However, Wi-Fi cards will need some additional parameters specified to manage their radio links to the Wi-Fi router or access point:

- **Mode** If your home network will be operating in ad hoc or wireless peer-to-peer mode, you won't have to worry about router settings. No router will be used since the workstations will talk directly with each other. All must be set to operate with a mode setting of ad hoc. If you are using a router to interconnect one or more Wi-Fi workstations, the mode setting for all should be Infrastructure.
- **SSID** Your system identifier is the name you have chosen for your local network. The router manufacturer will assign one at the factory, but you should change it to make your network harder to break into. The name should be the same for all the stations on your network. For the sake of consistency, remember that upper- and lowercase letters should be used in the same way for all. Your network name must be different than the names of any other Wi-Fi networks in the pickup range.
- **Encryption** The default is usually Disabled, which is fine for the brief period you will be establishing and setting up your network. Afterward, if you desire any privacy, you must turn this feature on in order to scramble your data stream for outsiders. When you do, you will be prompted to enter a numeric key or a pass phrase, which is a unique string of characters used to automatically generate a numeric key. This determines how the data stream is encoded and decoded. Obviously, if

it is not uniform for any of your workstations, they will be regarded as outsiders and unable to participate in your home network. Also, you will be asked to specify the level of encryption, usually 40-bit (sometimes called 64-bit) or 128-bit encryption. The more encryption, the slower the link. This encryption level also must be the same all around. Ad hoc mode is no different, except that no access point is involved. The WEP key must be uniformly configured on all your ad hoc wireless clients.

- **Channel** This is the frequency that your PCs' "walkie-talkies" will be using. Up to 14 are supported, but if you are in Europe, you will have to check to see which of them you can legally use. It must be the same channel for all stations. If someone else within range is using the same channel, you or that person must change to a different channel. As with the wired computers, your router will have an LED to show the presence of a wireless signal, and the remote wireless interfaces will have indicators as well (unless the interface is built in). These LEDs are the first and simplest indicators that the hardware is working or not. When any Wi-Fi interface is on the same channel as the router, the router's signal detect LED will light up.

To communicate, your wireless workstations and the router/access point must be configured identically, so you may want to determine the router's wireless settings at the outset and use them as the standard for the other stations in your network. If you use default values all around, then those parameters will probably be identical for all. But different manufacturers choose different defaults. The point of Wi-Fi standardization is the guarantee that all certified devices can be made to work together. But if you are mixing equipment, you will have to check and perhaps change some parameters in order to match them up. After this has been done, you should be able to see all the stations on your network from any of the wired or wireless PCs. (If you've installed a software firewall on the PCs, disable it until everything is working well.) Here's how to take attendance:

1. For Windows machines, double-click the Network Neighborhood icon.
2. You should see an icon bearing the name that you specified for your workgroup. Double-click it.
3. You should see the names of all the computers you've configured, named, and rebooted.

When your router is fully operational, if it has Universal Plug and Play enabled, it will also place an icon for itself in this display. If one of your stations is not on the list after a few minutes, check the Wi-Fi NIC's LED for a signal and recheck your parameters. If some of your workstations are OK, then try to determine what's different about the one(s) not visible.

Configure the Router

Although this task involves the same steps for workstations, it is more complex because today's residential routers are surprisingly flexible. This means that they offer many options and in turn many choices for you to make. Unlike workstations, which can usually be set up automatically with a manufacturer's CD-ROM, the router has no CD player of its own. Some vendors offer web-based utilities that will assist you, but you will have to log into the router first and do at least some configuration from your PC's keyboard.

Routers can be administered from any PC, including wireless ones and even those outside of your home, but you will have to weigh the convenience of that feature against the implied security threat. It's not wise to allow others to control your router. The safest way is to restrict access to one hard-wired terminal and configure your router from there.

Logging In for the First Time If you have been using your PC as a dial-up or have your browser configured for use in a corporate environment, you should check a few things before you use it to communicate with your local router. For Internet Explorer,

1. Under the Internet Options screen, click Connections.
2. Select LAN Settings. Verify that none of the following have been selected:
   - Automatically detect settings—no.
   - Use automatic configuration script—no.
   - Use a proxy server—no.

If your favorite browser is Netscape Navigator,

1. Click Edit. Then choose Preferences.
2. Click Advanced. Then select Proxies.
3. Choose Direct Connection to the Internet.

When you first activate your browser, if you have a favorite startup home page preset, the browser will attempt to connect to it as usual. But since you haven't yet instructed the router how to log into your ISP, no link can be made to the Internet and this connection attempt will fail. At this point in the process, that's normal. Point your browser to your router instead by entering the router's factory-assigned local IP address, such as http://192.168.2.1. A vendor can assign addresses from a range of non-routable IP addresses. The actual router address you use will be in your user manual, along with the factory-supplied login name and password. Enter those when prompted to do so. As with all passwords, remember that upper- and lowercase are important. If you are incorrect, you will receive a 401 (Authorization Required) error, and you must restart the login from the beginning. A successful login brings you to the router's main status screen, where you may begin to marvel at all the information that's contained in that small piece of hardware (see Figure 4-22).

The content of this screen and its submenus is not standardized. However, the same functions have to be addressed by all, and vendors tend to make common choices as to what should go where. As you can see, this main status screen gives you a quick overlook of the router's condition. It is the

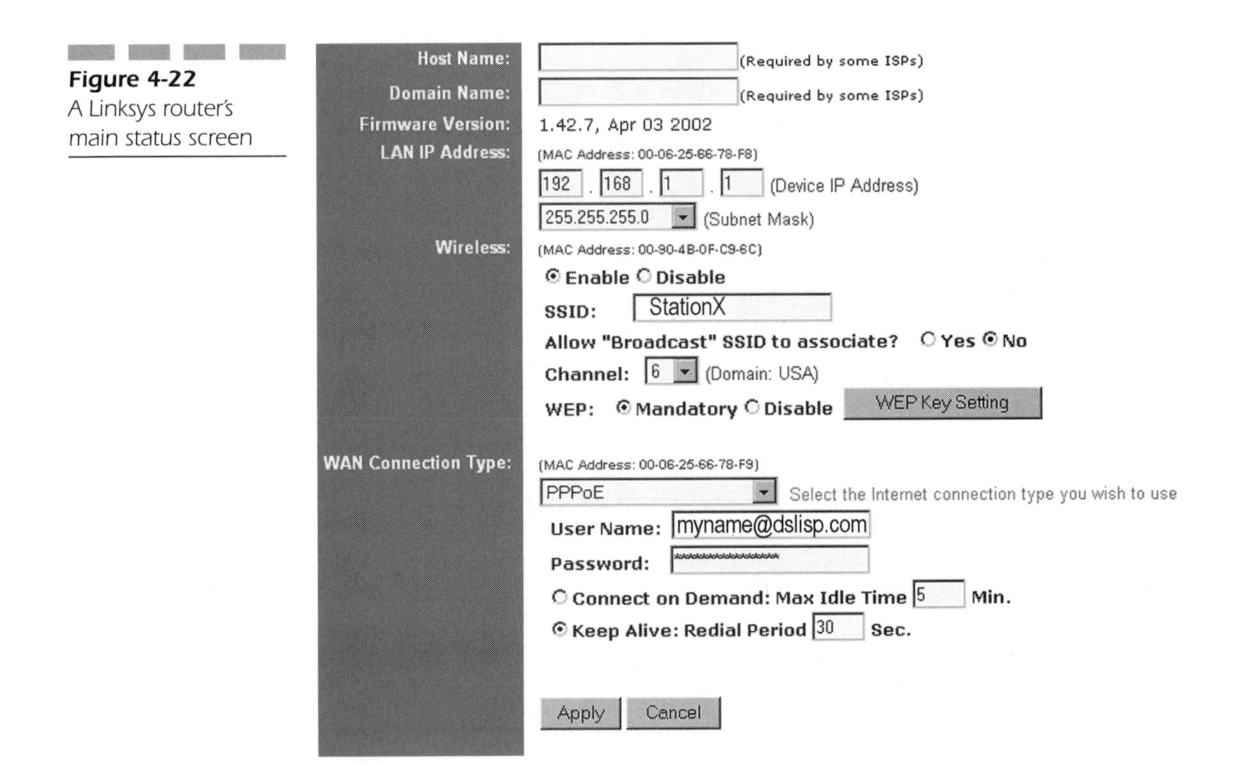

Figure 4-22 A Linksys router's main status screen

screen you will look at first and most often, particularly when troubleshooting, because it tells you whether or not the router has successfully logged into the Internet. The fact that you are seeing this screen at all is a hopeful sign that most of the router is functioning as it should.

Perhaps the first thing on your router for you to customize, after verifying that it seems to be working, is the login and password you just used. If the router let you in without a password, it is very important that you set one of your own. Until you do so, everyone else who has purchased that particular brand will know how to peer into yours. Whatever you change it to, write it down, perhaps on the configuration sheet you are using to record IP addresses and serial numbers. If you forget, the only way to regain admission will be to completely reset the router to factory defaults. This erases all the configuration changes you may make.

If you have other devices on your local network, such as additional Wi-Fi access points, they may have the same factory-assigned local IP address as your router. It's a simple matter to change it to another nonroutable, local address, but make sure you don't pick one that might be assigned to a workstation. Every device must have a unique IP address. Write your new choice down because your other workstations will need it in order to access your router.

Configuring Your Router for ISP Login This is the point at which the numbers you copied from your original standalone PC become useful. In order for your router to emulate the PC and log on to the Internet in its place, you will first have to select Static IP or PPPoE. If your address never changes, enter it along with the subnet mask and the ISP's router address. If your old PC logged into your ISP and received a different IP address every time, you must enter your ISP login name and password.

In either case, if your ISP's cable/DSL modem is expecting to see the MAC address of the original PC's NIC card, you will customize that now also. You may have to turn the power off and on in order for your changes to take effect. Afterward, it's easy to check whether or not your configuration was successful. The status screen will tell you if the router made the connection to your ISP and you can test all the router's wired client workstations PCs by directing their browsers to a familiar Internet web site (see Figure 4-22). (You may have to do a little more work to enable the wireless workstations to do the same.)

Proceed to enter the settings for SSID, channel, and encryption as you did for the workstations described previously in this chapter using the values you have determined as standard for all. (Some routers will enable you to establish multiple encryption keys. When you access the router from a

Figure 4-23 Red light, Green light. Some routers are clever enough to permit or deny access to the Web based on workstation IP address, on type of data, or even time of day.

IP	Port	Type	Block Time	Day ~	Time	Enabled
1. 192.168.2. ~	~	TCP / UDP	Always / Block	~		
2. 192.168.2. ~	~	TCP / UDP	Always / Block	~		
3. 192.168.2. ~	~	TCP / UDP	Always / Block	~		

Click ENTER to save settings and continue. ENTER

wireless station, that station must supply a pass phrase and the number of the key that the pair will be using from that point onward.) You may have to reboot your equipment for the changes to take effect. Afterward, you can test the wireless workstations on your Wi-Fi net by surfing them to a familiar Internet web page.

Connection Duration Cable/DSL modem Internet links are not automatically always on. PPPoE connections, like their telephone modem progenitors, can drop due to inactivity. You can determine if your router does so, and also the interval after which your router surrenders the connection (Maximum Idle Time). You can also choose whether or not your router will automatically log into your ISP again upon waking up due to activity demand (connect on demand). Since you will not be charged any more for it, you can also specify that the router will periodically send some packets to your ISP to hold the link open continuously. This is called the *keep-alive option*. It does not always work. Your ISP can force their equipment to drop you periodically for maintenance or other reasons. Unless you pay for a static IP address, your cable/DSL modem may be running with a new IP address overnight.

Special Application Configurations

For many users, the steps described previously will be all that is necessary to use a secure home Wi-Fi network, share information between local workstations, and share an Internet link among them. Some users, however, will have special applications and special configuration requirements. As explained earlier in this chapter, your Wi-Fi router makes your network more secure by erecting a firewall that blocks unrequested data coming from the Internet, but NAT does not automatically know which, if any, of your PCs should receive this data. Web servers, IP phones, chat rooms, or Internet gaming require that you prearrange holes in the NAT firewall (see Chapter 7). Most of the time it is possible to do so selectively and still keep your Wi-Fi PCs safe. The exemption goes by formal names such as *virtual server*, *port mapping*, or *port forwarding*.

Port Forwarding You've probably noticed that you can have multiple connections running to multiple Internet sources at the same time. For example, you can do software downloads from two sites while you surf a third and check your email. Your system sorts out all these incoming data streams by virtue of the software ports to which they are addressed. These are identification numbers inside the Internet packets, but they function as subaddresses that send particular data to particular programs within your PC. When you only had one PC connected to the Internet, they were of no consequence to you because all incoming traffic was destined for your one PC. But the purpose of a router is to connect several computers.

Routers can be programmed to send data destined for a particular port to a designated IP address on your home network. If all programs use the same ports for the same purpose, this would be easy to set up, but no industry-wide standardization exists. Different Internet computer games or chat programs, for example, will use different sets of ports. This means that you may have to do some experimentation to make your program work behind your firewall. Here are some rules to remember:

- **The target PC must have a static IP address.** If you decide to make one PC the recipient for data aimed at one set of ports, the PC has to have a static local IP address. If the router gives it an address at boot time, its address may change. The packets will head off toward the PC that bears the designated IP destination, whether it needs them or not.
- **There can be only one PC per port.** Only one computer inside your firewall can use a designated inbound port at a time. If you use multiple applications on multiple computers, and those programs expect data from the same port number, you will have to choose which PC owns the port. Either that or run both applications on the same PC.

- **Open only the doors you need.** You may wonder why you can't just open up all the ports and be done with it. If you did so, all the data coming to any port would go to one PC. For that PC, it would be the equivalent of turning off the firewall. Open ports are security risks.

For a practical example of all this, let's suppose that you want to set up your very own web server so that you can publish whatever you want on your own web site. You can get to your site from any of your home PCs by using its local IP address as the URL. But friends and relatives can't find it from outside when they use the cable/DSL modem's Internet IP address as the URL. When their browser requests arrive at your router, the router doesn't know what to do with them. It will not know which of your several PCs is the one running the web server until you tell it (see Figure 4-24).

Fortunately, for this example, web servers always use the same ports to communicate. Specifically, they are 80, 83, 1080, 8080, 8088, and 11523. We are sending any traffic coming in on those ports to the web server whose unchanging local IP address is 192.168.1.113. You can see another reason for not forwarding any unnecessary ports—you don't have an unlimited supply of choices. The ports listed here are all the ones that *may* be used.

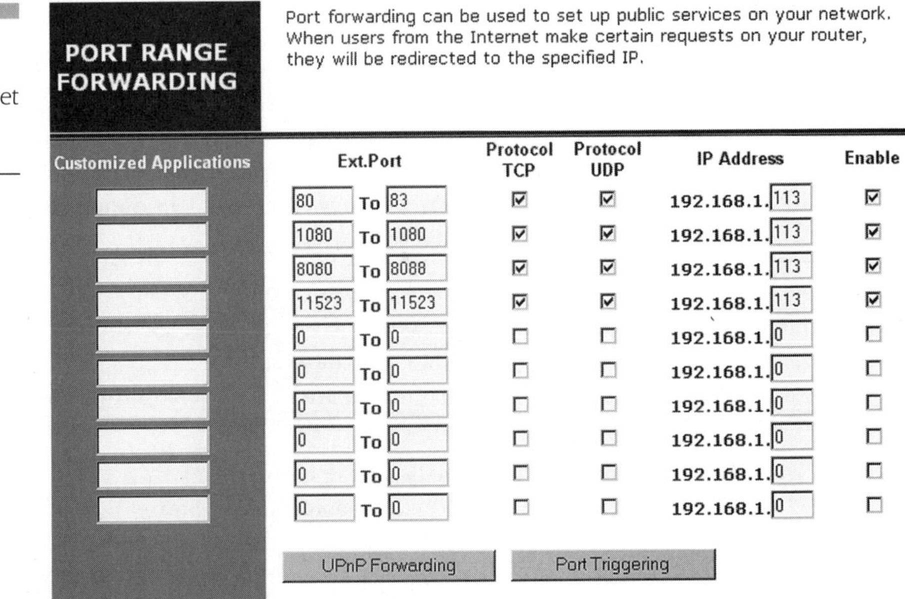

Figure 4-24
A Linksys port forwarding screen set to accommodate a home web server

Our web server works using only port 80 and port 21 for *File Transfer Protocol* (FTP) file transfers. It's easy to run tests by unclicking the enable box and then browsing your own by the WAN IP address of your cable/DSL modem (not the *local* web server IP address). If you open ports for Internet gaming, you may want to disable forwarding by unchecking those boxes when the game is over.

Trigger Ports As noted, some applications require groups of ports. If this happens in a predictable fashion, you can program your router to react to a single port request by opening the others as well for as long as the application is active. If you know, for example, that Deathbomb II will use port 1234 whenever the game begins to set up a connection to a game server, you can specify that as a *trigger port*. The router will forward the other necessary ports to a designated PC as long as 1234 is in use.

The Demilitarized Zone (DMZ) Some applications such as teleconferencing or IP telephony negotiate ports automatically when one user connects to another. It is therefore impossible to guess which ports the two connection robots will agree on. These applications do offer a method of last resort, however. You can put that local PC outside the firewall, or put another way, inside the *demilitarized zone* (DMZ). When you have done so, you do not need to mention the PC's IP address in port forwarding. All the ports will get forwarded to it. The PC involved should have its own personal firewall and antivirus protection installed, as it is exposed to the Internet as though there were no router. Routers only enable one total exemption of this nature.

If you are trying to get a special-application PC to work, you might put it in the DMZ temporarily as a troubleshooting measure. If it works there but not otherwise, you know that port forwarding is part of your connection problem.

Access Controls

MAC filtering is an effective method for keeping outsiders from using your Wi-Fi router. When it is enabled, this feature examines the source hardware address of every incoming data packet and compares it to a list of addresses that have been specifically preadmitted (see Figure 4-25). As described earlier in this chapter, the MAC address of every Wi-Fi interface is unique (unless a hacker has managed to clone one of yours) and consists of a manufacturer's ID plus the interface's serial number.

The ways in which you can set access rights for your network's computers can be very detailed. You, as your own system administrator, can decide if unannounced users are allowed to access anything at all—or everything. Some brands allow you to determine these parameters separately for hardwired PCs (Wired Connection Control) and for your Wi-Fi PCs (Wireless Association Control).

Client Filters In addition to totally denying access to the router, some routers are flexible enough to turn access on and off for given applications at given hours of the day. Here are some things you can do with these refinements:

- Make sure an employee can't surf the net during working hours, but can copy files from local PCs. After 5 P.M., you can use the same PC as you wish.
- Turn off game port forwarding on your kids' computers and back on during weekends.
- Deny access to your web server during scheduled maintenance.
- Allow a customer in your living room to copy a PowerPoint presentation from a server, but not send email.
- Allow your kid's pals to get to the Internet, but not to the PCs you use for your home business.

If you do all these things at once, it can quickly become complicated. Don't be surprised if like most system administrators, you wind up facing a line of users complaining about their restrictions.The client filter screen is the one you would use to allow for VPN, or Virtual Private Networking, if your router is capable of it. You can enable or disable passthrough for the IPsec and *Point-to-Point Tunneling* (PPTP) protocols.

Date and Time If your router is clever enough to adjust to the time of day as mentioned in "Client Filters," then it must have a means for knowing the time. Most time-sensitive routers can query any of a number of public Internet servers for this information. They periodically broadcast a *Simple Network Time Protocol* (SNTP) request. They can set their own time more accurately than you can from the keyboard. They don't know where they are, however, which means that you will have to enter the time zone in which you reside.

Assorted Router Utilities

Today's routers do much more than route packets from one port to another. Manufacturers have built-in utilities that make them easier to control and understand. Not all home routers will boast of all the following value-added features, but most will. Here are some brief descriptions of them along with the ways you can make them work for you.

Logging Options Routers can tell you a great deal about what they see, and at least in the beginning, these tales can be very educational. Some may only keep a record of attempts to access your network from the outside. Others can keep detailed records of inward and outbound traffic for each of the workstations on your network.

Routers can track the source of incoming packets, too. If you have a local web server, your router may be able to tell you the IP addresses of everyone who has visited your site. If this feature is enabled, you will probably have to specify the IP address of the workstation you will allow to view the logs.

Router Routing Although you may have more than one wireless access point in a large home, you would probably need a house bigger than the Playboy mansion to justify more than one home router. Even so, some home routers have enough intelligence to carry on a conversation with other routers on a network to determine the fastest path from one end to another. Like their big brothers in the data centers, they negotiate using the *Router Information Protocol* (RIP). You can also manually specify the way that data packets should flow between multiple routers, but your router should almost always be running as a single-point gateway, which is the default.

DHCP Controls DHCP is the method by which your router can hand out IP addresses and other parameters to your workstations as they boot up. If enabled, it is automatic, but you retain a great deal of control over the process. A prime security concern is the exclusion of foreigners, who might be tempted to hitch a ride on your Internet connection. If you know the number of users on your network, you can restrict the size of the available IP address pool so that your own PCs are taken care of but no others.

You can also determine the starting point for the range of IP addresses that will be assigned. You can pick something other than the defaults in order to make your IP addresses harder for an outsider to guess at.

Remote Router Management This is the screen where you choose the terminals from which your router is configured. You can allow configuration changes from anywhere on your home network or even from the Internet, but to do so would be asking for strangers to take ownership of your router. Unless a compelling reason exists to reconfigure your router when you are out of town, you are better off if you enter a local IP address for a wired PC in your home, preferably one that is also password protected in some way. Any home user whose access has been restricted will have a motive for finding his or her way into your home router.

Software Reset Sooner or later your Wi-Fi router may become dazed and confused by noise on the networks, by static electricity shocks, or for a variety of reasons. Multiple ways are available for bringing a router to its senses when it locks up. Don't be too hasty to push the hardware-reset button, because that clears everything out of the router's memory, including all the preferences that you have painstakingly entered. You can even avoid a jarring power fail reset if your unit comes with a software-reset option. This will clear temporary memory and communication buffers, and restart the internal programming, but it will leave your passwords and other semipermanent data intact. It is the router equivalent to a computer's warm boot.

Depending on the brand of your router, the power and port lights may flash to indicate that the reboot is in progress and startup diagnostics are running. Some brands let you perform a warm router boot by pushing the reset button for just a second, and then a total reset if the button is held down for several seconds. Determine how your router will react before you push the button.

Restoring Defaults This option is the same as pushing the hardware reset button, except that you initiate the process through a keyboard instead of a push button. Afterward, all the router's parameters will be the same as they were when you first pulled your router out of the carton it came in. Use it as a last resort, because recovering from it may be labor intensive.

Updating Firmware As with PCs, it sometimes becomes necessary to improve your Wi-Fi router's operating software in some way. You may need to do this in order to fix a newly discovered bug, counter a new kind of hacker, or add some new features that you would otherwise get only with new hardware. Typically, you would do this when requested by your router manufacturer's technical support team or in response to a notice on their web site. The main status screen will tell you the firmware version that your router is currently running. Look for a higher version number (or a later date) of the software on the web site, and be certain that the file you choose is for your particular model. If you are mistaken, you might have to send the unit in for repairs.

The actual process is usually very simple. You enter a password and the name of the file you seek, and sit back. Your router is clever enough to log into the vendor's web site and copy the new software across the Internet, just as you periodically do with your computer. Some models will accept a compressed patch file from one of your workstations after you manually copy it there.

Some models will let you trigger the upgrade process from across the Internet, but you are better off doing it from a close vantage point so that you can observe the indicator lights and restore your user preferences afterward. Often, those parameters are lost when you reprogram your router, just as though you had pushed the hardware-reset button. Write them down before you start.

In the first half of this chapter, we took a hard look at all the possible outside threats that must be considered when you network your home. Not all of these will present themselves everyday, and if you start out by assuming responsibility for your network and preparing it properly, the threats won't amount to more than an occasional chore. In the second half, we detailed how to set up Wi-Fi routers and workstations with an eye toward system security.

The next chapter deals with threats from within. After you have made your home Wi-Fi network into a major communications channel for your family and a tool for school or work, you won't want it laid low by flickering ceiling lights or distant thunderstorms.

CHAPTER 5

Good Housekeeping

> Everything put together by the hand of man falls apart eventually.
>
> —Simon and Garfunkel

This chapter discusses the costs and means of protection from environmental problems such as static electricity, bad power, and bad luck.

Maintenance Is a Good Habit

It's a busy world we live in and that makes it difficult to spend any time that does not seem to be immediately profitable. That's why setting up a regular schedule for keeping your home network in shape is important, and whatever schedule you choose for backups, it's a good idea to stick to it. Friday seems to be a good day for a weekly routine. (It's a good way to look busy while waiting for the weekend to arrive.) Our routine includes

- A partial backup, at least, to a server.
- Dump the recycle bins and run a disk cleanup program if available.
- Refresh the virus definitions and run the scans.
- Similarly, refresh parental control lists, if you use that software, and the firewall lists.
- Run a disk error scan and perhaps defragment the hard drives as well.
- Check web sites for new patches for your PC, your browser, and your *wireless fidelity* (Wi-Fi) equipment.
- Tell your browser to delete accumulated cookies or run a spyware scan.
- If you work from home, erase deleted files beyond recovery.

We'll go through all of these in more detail in the sections that follow. Some of these activities, such as disk defragmentation, will require the full resources of the PC. Others, like the virus scan, can proceed minimized in the background while you do something else.

Scout's Motto: Be Prepared

It's true that some blessed individuals sell their used cars without ever having pulled the spare tire from the trunk, but practically everyone else needs one sooner or later, and the same could be said for fire extinguishers or health insurance. It's unpleasant to contemplate, but you will only appreciate all the man-hours you've sunk into your personal computer when the accumulated work disappears. You might try a little experiment yourself. Turn your computer off, pretend that it is never coming back, and try to write down everything that was on it. Then turn it back on and compare its actual contents with your list. Here are some real-life scenarios that illustrate common dangers:

- You have several duplicate folders open on multiple home PCs over your Wi-Fi link. To clean house, you delete the duplicates. After a few moments, you realize you can't remember which were which. When you remotely delete a file, it doesn't go into a recycle bin.
- You read file after file with no problem and then discover that you cannot read any of them a second time. As you ponder this mystery, you notice that your hard drive sounds like a knife sharpener.
- After an overnight thunderstorm, all the digital clocks in the house are blinking. The PC's screen, however, shows a steady "blue screen of death."
- You add a larger hard drive to your PC and copy the old drive's contents into the empty new one, but the drive jumpers were incorrect. You now have two copies of a blank drive.
- When you upgraded to the latest operating system, a popup window told you to make a backup first, but you were anxious to get going and you never had a problem before. Now it won't boot.
- Your kid walks across the nylon rug wearing his sneakers. When he touches the keyboard, you can hear the *snap!* from another room. Afterward, the PC complains about a file allocation table error.
- You email your customer that you are dropping him because he hasn't paid his bill after several warnings. You angrily erase his folder and empty the recycle bin, even as his check sits waiting in your mailbox.

You can be as sensible and diligent as you like. Your data will remain vulnerable. The best way to withstand a data catastrophe, of course, is to avoid it. This chapter will discuss prudent steps you can take that will narrow the odds of an ugly surprise, but those odds can never be reduced to zero.

Call for Backup

If you work from home, a hard disk crash can be as devastating to your business as a runaway house fire. If you have a copy of your data, even one that's months old, the disaster is reduced to the scale of a survivable nuisance. If nothing else, the aged backup will give you a clue as to what will need to be updated after the recovery. That's why making copies is common sense. Fortunately, the Windows operating systems make backing up easy to do, and Wi-Fi connections can make it possible from anywhere in your house.

Backup Software

Emergency copies are so important that the Windows operating system includes bundled software for assistance. This software is not always copied in when your system is first installed, but installing it later is not hard to do. If you add a tape backup system, the manufacturer will provide bundled software to control the tape drive and manage backups to it. To find out if you already have Windows backup software on your computer, simply

1. Double-click to open up the desktop My Computer icon.
2. Right-click the icon for your hard drive.

In the window that opens, "Backup" will be presented as a choice. If you don't have the option, you can still back up any file on your PC by copying it whole to media on other PCs. The software gives you some powerful options (see Figure 5-1). With it, you can

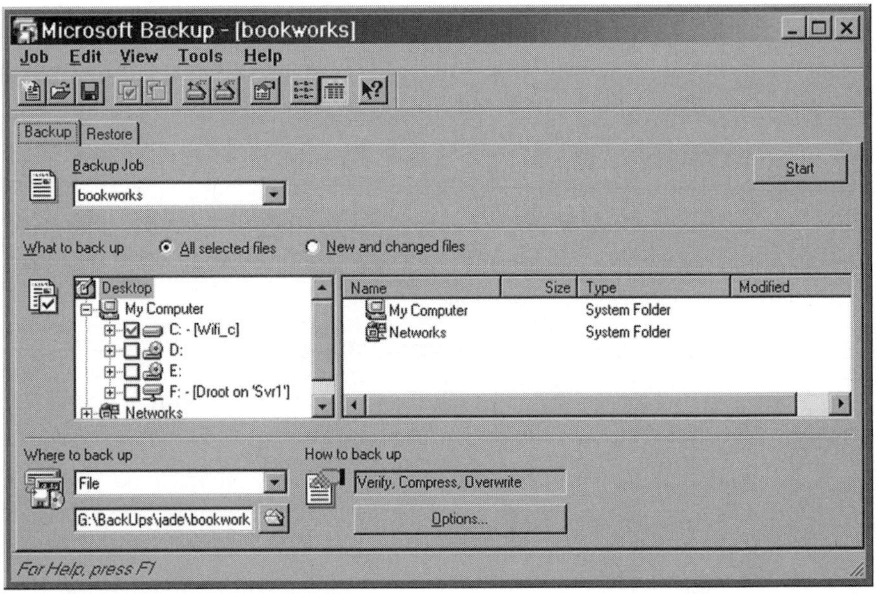

Figure 5-1 The Windows backup utility provides many options for insuring your files.

- Restore the files in the same groups you chose for their backup.
- Determine which folders will get backed up to which repositories.
- Keep logs of the results for later viewing.
- Compress the files as they are copied over to save space and choose the amount of compression.
- Back up system files that you otherwise could not access.

Given the small amount of disk space required to load the program (5 MB), having this facility on your PC is a good investment. In earlier versions of Windows, it was available as an optional component located in the System Tools folder. Unfortunately, earlier versions of the Windows task scheduler could not be used to run a backup unattended. Later versions such as Windows 2000 will. Starting with Windows ME, the backup installer program is found on the installation CD. Here's how to copy it in:

1. Put the CD in the drive. When it spins up, you will see a menu screen.
2. Choose Browse this CD. Double-click the folder labeled Add-ons.
3. Open the subfolder named MSBACKUP. Click to run the install program MSBEXP.EXE.

The installer will give you the rest of the loading instructions. After installation, to launch the program

1. Start at Start, select Programs, and choose Accessories. Then click System Tools.
2. Click the Backup icon to launch backup.

When backup runs for the first time, it will search your system for dedicated backup devices (primarily tape drives). If the program cannot detect a device that actually exists, you will be asked whether or not you want to run the Hardware Install wizard. If you have been in the good habit of running backups using a previous version of Windows, then installing this version of backup will not overwrite the old one.

Tape Drives The backup utility assumes that you will be backing up to a tape drive. (That's what it was originally written for.) One advantage of wireless networking is that you can redirect copies to any of your hard drives; particularly those on another networked PC. Likewise, you can back up files from any PC across your network to a shared tape drive. If the files you want to back up are small enough, you can even back them up to a floppy disk. If you do happen to use a tape drive, how you proceed will

Figure 5-2
A built-in backup tape drive and one of the cartridges that it uses. Photo by Daly Road Graphics.

largely be determined by the drive's hardware capabilities. An older cartridge backup tape such as the Colorado line (see Figure 5-2) will hold only 3GB of compressed data. Newer models using the dds-3 or *digital audio tape* (DAT) recording technology can squeeze up to 24GB onto a smaller, cheaper tape cassette. DAT-40 drives can accept up to 40GB. A backup to one of these can take hours, and if your system is large enough, you might have to manually change the tapes from time to time.

Tapes have other disadvantages. They deposit their recording medium, ferrous oxide (refined rust), on the drives' recording heads, which eventually must be cleaned with a cotton swab. If you back up every night, you will eventually have to replace the pickup heads because, as with any moving part, friction wears them down.

Tapes do have the advantage of giving you a transportable medium, sometimes referred to as a *bank copy*, which can be stored completely offsite. That way your most valuable data can withstand even an earthquake. Some come with "one-button" firmware recovery programs that promise to reconstruct your damaged file system with ease. Manufacturers of these tape drives and cassettes usually include software packages that let you schedule backups for low usage hours. Windows also has a built-in facility that helps to handle details, such as what to back up and what to ignore. These customized backup lists can be stored by easy-to-remember names, such as Thursday_Partial. The data can be encrypted on the fly and stored with a password. This is a good idea if you are dealing with business-related information. The backup programs also store reports about how the jobs went or failed. Checking these periodically is a good overall diagnostic tool.

Copying to CD-ROMs Modern CD-ROM drives have the capability to record as well as read. Unfortunately, their maximum capacity of 700MB is not sufficient except for nightly, partial updates. The cost of blank, recordable CDs has dropped to $.50 each in bulk. If you use a drive model designated as *CD- rewritable disk* (CD-RW), you can use a single rewritable CD night after night. Remember that rewritable CDs can only be read from rewritable CD drives. Recovering data from them onto a different PC with a read-only CD drive will present you with a problem. However, it is possible to share a CD-ROM across a (Wi-Fi) network just like any other hard drive. We detail how to share such devices across a network in Chapter 4's "File Sharing."

Copying to Another Workstation Because the cost of disk storage has fallen, using a hard drive for backups has become faster and easier than using tapes, although if you have a tape drive there's no reason you cannot use both media as you deem appropriate. A home Wi-Fi network enables you to easily make a straight, exact copy of files from one PC to another, and this might turn out to be the best way for small collections of files. One feature that will help in this regard is backup's compression-on-the-fly feature, which typically puts 1.5GB of raw data into 1GB of backup file space. This amount can vary, depending on what type of files you are copying. If a file has already been compressed, such as a .jpg, .gif, or .mp3, then compressing it a second time will actually make it a bit larger. Some operating systems limit the size of a single file to 5GB, which may not be enough for everything you want to protect. You can work around the backup file size limit, along with your own time constraints, by being selective in what you back up and how often.

When to Back Up You don't have to copy everything on your system because everything does not change every day. Operating system files, for example, only change when you install upgrades or patches. You can pick out the directories in which these reside (\Windows, for example) and back those up, say, once a month. Directories that do contain dynamically changing data such as \My Documents or email folders may need to be backed up daily. As mentioned, with later versions of Windows and some versions of vendor-supplied software, you can schedule these to occur overnight. You can even run them while you use the system, although you cannot back up a document that is actively being used by the system.

Registry Backup Your Windows PC keeps a large amount of system-crucial information in its registry. It stores system configurations, including user account information and preferences, network protocol bindings, and runtime settings for every program you run. That's why you can't simply copy programs from one PC to another and expect them to work. The registry is so important that your computer automatically maintains a copy of it in addition to any you may choose to make on your own.

Whenever you boot your PC, a registry diagnostic program automatically scans the file. If problems are discovered, the program will automatically replace the damaged registry with the backup copy. You can start this diagnostic yourself and make a backup copy whenever you want by following these steps:

1. Begin by clicking Start.
2. Open Programs. From there, open up Accessories.
3. Select System Tools, and then click System Information.
4. In System Information, click to select the Tools menu.
5. Click Registry Checker.

Network Backup Repositories

Making a spare copy on the same PC is a waste of time, because if that PC dies you've lost the copy as well. You are better off isolating the copies from the source. MIS professionals copy essential corporate data onto removable media, some of which is periodically taken away and stored offsite. These are called *bank copies* because they are sometimes stored in armored vaults. This used to be considered routine drudgework, but after the 9-11 attacks the task is taken very seriously.

With a Wi-Fi net in place, you can cross-copy between all your machines, but it will help you stay organized if you create a directory on one backup server specifically labeled for backups. Multiple backups on multiple machines will quickly become mind-boggling, and duplicate copies are a waste of hard drive space. An ideal backup server will have lots of storage, but it need not be very fast because you won't need it often. An older machine might be best to use as your "bank vault," perhaps one that you are also using as a software router or gateway. Throughput does make a difference for this task, which involves massive amounts of data moving at a constant, sustained rate. Your best PC will be one that's hardwired to your Wi-Fi router through one of its 100 Mbps switched ports, perhaps the same one you used to securely configure your router.

A Backup Speed Run Once your backup jobs have been defined, running them is not much of a bother. You can continue to work with the backup window minimized. Your cursor may seem a bit sticky at times, but unless you are doing some high-demand, real-time task, you won't notice. Table 5-1 shows the results of some backup tests to give you an idea of how long these tasks are.

As you can see, a full-demand backup is one of the rare instances in which the speed differential between wired and wireless becomes obvious. You can make it less so by specifying that only new or unchanged files be added to the backup after you initially perform a full one. In the previous example, for instance, of the 3GB to be backed up, only 16MB were actually new material because not many files change over time. You are also given choices as to the amount of compression applied to the files on the fly. Less compression means the backup will proceed faster, but take up more space when completed.

Within a single designated backup directory, you may create named subfolders for each of your machines. You may save each of your backup lists as named jobs, so make the names descriptive and include the source machine's name, such as JoeLapWeeklyDocs. That will help you retrieve the backup—it's worthless if you can't find it—and also keep it separated from others. If your vault does not have enough space, Wi-Fi makes it easy to shuffle files around to other machines to free up space. You can also easily check out the contents of other PCs' hard drives to see if you've unnecessarily duplicated files that can be shared instead. This is one example of how your Wi-Fi network can become greater than the sum of its parts.

After you've determined your hardware and your method, test it at least once by recovering it, perhaps to a temporary directory on another PC. Consider it as a fire drill. Don't wait until you absolutely, positively need to retrieve an item to find out that it wasn't included in the backup list or that

Table 5-1
Backup Tests

	100 Mbps Wired	11 Mbps USB Wi-Fi
Elapsed time	28 minutes	126 minutes
Processed files	6,257	6,304
Total bytes	3,183,241,943	3,198,871,604
MB/minute	113.6	25.4

you've been backing up a different file with the same name from the wrong folder.

You may wonder, can you back up the backup PC as well? If you have the unused space for it, you can copy anything. But as your own SysAdmin, you will have to make your own cost/benefit analysis about all these measures. No amount of time or money will make you completely safe, but a small amount will greatly decrease your overall risk.

Power Corrupts

Although it is true that none of this equipment will run without power, it is also sadly true that power can kill data equipment. Transistors replaced vacuum tubes in the first place because transistors operated at much lower voltages. The microscopic transistors packed by millions into the space of a postage stamp are very susceptible to high voltages. A laptop insurance agency, safeware.com, says that in 2001 over 59,000 laptops were lost to power surges. (They don't track losses to desktop PCs.) The power supply in your PC will automatically adjust for varying house power, but if the AC voltage drops below a certain level, a condition known as a *brownout*, the power supply won't have enough to work with. If the PC happened to be writing data to the hard drive when the system died, that data may be corrupt.

Computer power supplies are also unable to cope with severe and sudden jumps in voltage, called *spikes*. A common cause is lightning, which does not have to actually strike a power line to do considerable damage. Any nearby wire will pick up a portion of the discharge and carry it, sometimes for miles. Owners of short-wave radios with large antennas are aware of this danger. Computer users with equipment tied to small antennas or telephone lines may not be.

Most spikes are small and inconsequential. Every time you turn a light bulb on or off, you inject one into the power cord. But a severe spike may dissipate itself through several pieces of attached equipment, damaging them all, or skipping some randomly, or possibly even causing damage that does not kill the equipment until months later.

Residents of California were surprised in 2001 when large portions of the state were suddenly forced to ration electricity. For many, their home voltage dropped to zero and stayed there for an hour at a time. Although the state authorities shut down some portions of the power grid and restored others, voltage sags and spikes rebounded throughout, even past the state's borders. These *transients* occur constantly from normal power company operations, and they happen everywhere. Your electricity may be relatively clean now, but even if you don't live in power-starved areas, it is prudent to go by the old adage that "What goes around, comes around." You should utilize equipment to protect your Wi-Fi investment. Remember that today's home entertainment systems also rely on transistors, and even a modern coffeepot has a microprocessor inside.

Not all spikes ride in on power lines. Any long wire can pick up electrical noise, even at a distance (that's how antennas work). Your telephone lines are equally susceptible, even though they are fused at the point where they enter your house. Surge suppressors are available for phone lines also, since many kinds of delicate electronic equipment, such as faxes, answering machines, and telephones, are connected directly to them. The telephone line protection built into the *uninterruptable power supply* (UPS) is an example. Unfortunately, some of these spike killers also throw a wet blanket on the incoming *digital subscriber line* (DSL) data signal, but specially made suppressors are available that do not. Radio Shack carries one for $15, (model number 61-2146.)

Wireless Depends on Correct Wiring The first step to safe power is a quick survey of your electrical outlets. All computer equipment now requires three-pronged, grounded outlets and cords to comply with electrical codes. All the protective equipment described here is designed with that assumption. In most cases, they bleed excess power into the third ground line. Professional technicians will always put on wrist straps that will connect or *ground* them to the chassis of the device they are working on for just this purpose. The simple act of sliding off a fabric chair can impart a memory-frying static electricity charge in dry weather. But if the outlet and chassis are not really grounded, then such measures are useless.

Unfortunately, you can't be sure that an electrical outlet is correctly wired just by counting the holes. Also, two- to three-prong adapter plugs and two-wire extension cords are sometimes mistakenly used for computers. You can buy simple test devices at hardware stores for a few dollars that plug into wall outlets (see Figure 5-3). Their indicator lights tell you right from wrong, and if wrong, why.

We were shocked (no pun intended) to find that several three-pronged outlets in our home were not grounded. The previous owners had replaced defective two-prong outlets with newer three-pronged ones, but never ran the third ground wire to them. Fixing that required a visit by an electrician.

We loaned the outlet-tester to a friend who lived in a recently built apartment complex. A check of his outlets revealed that all had been grounded, but in one of them the hot and neutral wires had been reversed.

Figure 5-3 The indicator lights tell you at a glance if your power outlet is miswired. Photo by Daly Road Graphics.

The investment in the little tester was worth the money, because a miswired outlet is unsafe to people as well as computers, especially in the kitchen or bathroom where water is present. Sending a jolt to ground through a wire is preferable to having one pass through your own two feet.

Ironing out the Volts

Almost all Wi-Fi hardware is powered through low-voltage, step-down transformers that isolate hardware from direct contact with AC lines. They operate by induction and dampen incoming voltage spikes, but no transformer can protect from an overwhelming nearby lightning hit, and you must protect other sensitive equipment as well. The simplest devices for diverting an incoming spike from your network can be found in hardware, computer, and electronics stores. *Surge suppressors* have a high resistance to the flow of electricity at the normal 117-volt level, but will resist it when it varies to a high voltage (thus the name *varistor*). Another method of achieving the same end is the use of a small neon bulb or other device to pass a spike over from one wire to the return wire in a deliberate short circuit. These also pass high voltages and so make an ideal on-off indicator lamp in multiple power outlets. Because of their simplicity and low cost, these types of components are often built into PC power supplies at the factory.

It should go without saying that the cheaper your surge suppressor is, the less effective it is likely to be. The plastic $8 to $25 models were not engineered with the needs of high-speed networks in mind. Given the cost of repair, including the man-hours involved, you may want to consider an industrial-grade device, one designed for high-speed data (see Figure 5-4). These suppressors can sell for $40 to $100 or more, depending on their current-carrying capacity and effectiveness. Some also include RJ-11 or telephone sockets, so that you can filter out spikes that ride in on your telephone lines. These spikes can be just as damaging as those on the power lines. Check the label. It should say that the suppressor is *UL 1449 listed*. Some manufacturers may offer an insurance policy to cover the cost of repair if their device fails, but if your data is gone, that will be cold comfort. Your best insurance policy is the price tag and the UL listing.

The next step up from a simple spike killer is a model that includes *chokes* in the lines. These pass electricity at its normal frequency of 60 cycles per second, but increasingly resist or absorb it at higher frequencies,

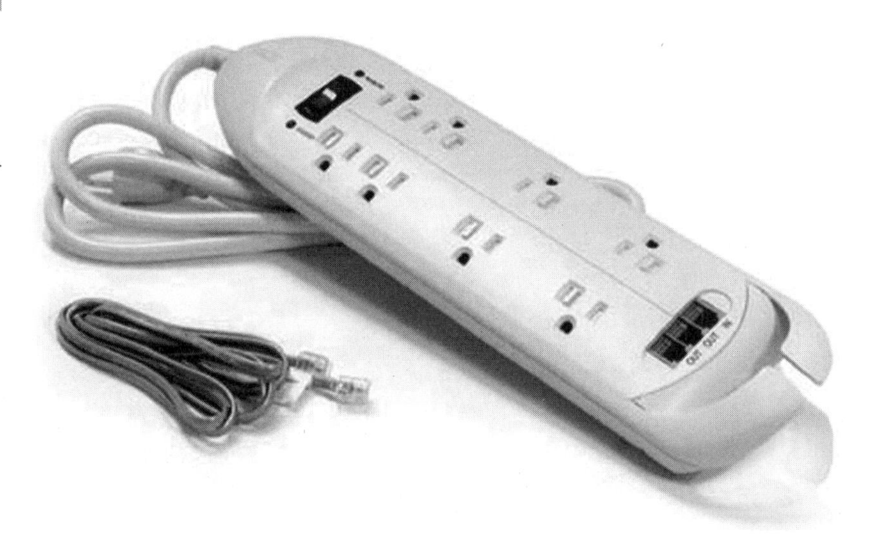

Figure 5-4
A heavy-duty surge/spike suppressor. Photo courtesy of Stanley.

especially radio frequencies. These resist spikes as well. Often a series of chokes and capacitors will be used to daisy chain one set of electrical outlets to the next inside the multioutlet box, so that protection accumulates from outlet to outlet. In this case, the outlets farthest from the power cord are the ones into which your most sensitive or most valuable equipment should be plugged.

Not all power pollution is deadly. Some kinds are just a nuisance, particularly for Wi-Fi computer networks, because power lines sometimes conduct *radio interference* (RFI). This electrical noise is defined by the Federal Standard *Glossary of Telecommunications Terms* as any electromagnetic disturbance that interrupts, obstructs, or otherwise degrades or limits the effective performance of electronics/electrical equipment. Possible power line noisemakers include

- Switching power supplies, such as laptop “power bricks”
- Nearby AM radio stations
- Fluorescent or mercury lights and halogen light power supplies

A Powerful Mystery

“We had one case where the net connection went dead every evening at seven and stayed dead until seven the next morning. Well, not exactly as those times. We worked with the customer day after day, long enough to determine that the outage period got longer by two minutes a day. It had us baffled. We figured the cause was environmental, probably something local, but we couldn’t prove it. Finally, the light bulb came on over the guy’s head—literally. He was staring out his front window at 7 A.M. when the streetlight in front of his house turned itself off automatically. He tested his LAN. It worked. He waited until seven that night and proved that his connection dropped immediately after the street light automatically turned itself on again. The power lines to the guy’s house ran right under that mercury vapor light. They were picking up noise from the light’s ballast. You know, seems like it always turns out to be something simple—after you find it.”

—Matt H., a diligent trouble desk technician for the nation’s second largest ISP

- Cordless phones
- Lighting dimmers
- Touch on-off switches

The last two can induce noise even when they appear to be turned off. Good suppressors filter most of the noise out. That's a good thing because line noise can slow down the throughput of your Wi-Fi network or stop it entirely (see "A Powerful Mystery"). Times will occur when a radio signal must intentionally pass through AC wires, such as the obvious case of a home network that uses power wires as the transmission media. Room-to-room "wireless" intercoms and computer-controlled remote AC switches send signals this way. If you plug them into a noise-choking power strip, they will be safe, but deaf and dumb.

A common user mistake is to plug a *source* of noise or spikes directly into a surge suppressor. One notorious example was the employee who plugged his coffee maker into the same power strip outlet that his PC used. This protected the rest of the building from the coffeepot, but killed his hard drive again and again.

Uninterruptible Power Supplies (UPSs)

UPS has nothing to do with package delivery when used to describe a power source. Strictly speaking, no truly permanent power source exists, but these devices draw on batteries to maintain power temporarily when the incoming AC sags or is blacked out entirely. Years ago these power backups were impractical for home use, chiefly because their batteries were heavy and dangerous. Some companies' requirements for backup power are so great that they rely on standby jet turbine generators, which most suburbanites cannot afford. But as small home offices have grown in number, new UPS units have been tailored to meet the need. They are smarter and more compact (see Figure 5-5). Instead of car batteries, they use built-in gel cells that do not spill acid. Some also include RJ-11 telephone sockets for your incoming phone lines to suppress spikes from that source.

As an added convenience, most UPS units provide additional AC plugs that suppress surges but do not provide full-time power. Use them to protect electronic devices that do not have to remain running continuously. When you are buying, check to see that all the outlets are spaced far enough

Figure 5-5 Above: The UPS can be controlled from the PC screen and vice versa. Below: A small UPS with additional sockets for phone line spike suppression. Photo by Daly Road Graphics.

apart to accommodate fat plug-in power transformers side by side. The UPS outlets that do give AC power for as long as the batteries hold up are almost entirely immune to spikes and noise.

Automatic Voltage Regulation If your model is truly uninterruptible, it will switch over to reserve power without pausing. Top-of-the-line UPSs are referred to as *automatic voltage regulators* (AVRs). They boost low incoming voltage or trim high voltage to a safe 110/120 volts by drawing and converting battery power whenever AC drops below 85 volts or goes over 150 volts. Some models will even alert an attached computer through a serial port cable that a power emergency has occurred. After that, the UPS will report on how much reserve remains at intervals you determine, so that a properly programmed computer can shut itself down in a safe and orderly fashion. You can also use the software control feature to preschedule an unattended power down.

UPS Duration Versus Loading The capacity of a home-office UPS is roughly determined by its price ($40 to $300), its size and weight, and the number and type of items you plug into it. You should protect your DSL modem, your wired hubs, and Wi-Fi access points foremost. Everything else depends on them. If they stay up, you can continue to work on a battery-powered Wi-Fi laptop by candlelight, if necessary. You probably should include a desktop PC if you use it as a gateway or server.

You do not need to include a PC's video tube display unless you intend to keep working in the dark, because unlike laptop displays, video tubes use up lots of power. You definitely do not want to plug in an air conditioner or coffeepot. The UPSs' overload circuit breaker would not tolerate such a load, and even if it did, the battery would be drained in a moment. Some users follow these commonsense rules and then absent-mindedly plug a multi-outlet strip into the UPS (which incidentally is against electrical codes). They soon forget the source of that strip's power and then plug unnecessary high-drain devices into it.

A UPS can be more than a business expense. It might prove very useful as a temporary power source for you or your family in the event of a storm or other emergency outage. If you want to include an emergency light, invest in a fluorescent model because they produce more light with less power than the cheaper incandescent bulbs. You may also want to plug in a cheap transistor radio tuned to a local news station. You can also use it to recharge cell phones.

Calculating Capacity Needs It is possible to predict the longevity of a UPS by calculating the loads and adding them up. Some vendors provide programs for this on their web sites as an aid to deciding which model you'll need. Having excess capacity is better than having to buy a second one. When you get it home and installed, you should take the time to test it under an actual load to assure yourself you've chosen correctly. Testing is also a good way to learn what the UPS sounds like when it's working.

Turn on the devices you want to support, making sure that any PCs among them are idling, as they do when the *basic input/output system* (BIOS) setup screen is active or when they are waiting for a login response. Then unplug the UPS from its wall socket, sit back, and watch the clock. Most UPSs will make some kind of audio alert to indicate that they are running on batteries, and they may begin to beep more frantically as they near the end of their reserves. Those with serial data output will count down their reserves on a PC screen second by second. The alert can also tell you when you've inadvertently turned off power at an electrical panel by popping a circuit breaker. It's a good idea to repeat the endurance test at least once per year, because the batteries inside a UPS will not last forever, and the devices you've attached to it may have changed in that time. Two to three years is a typical life span for a gel cell. Some UPS suppliers offer prepaid plans for periodic battery replacement.

Other Maintenance Tasks

Providing web access to all your home computers is something of a double-edged sword. Your need to keep your protective software up-to-date becomes urgent, because the source of potential problems is the global Internet and your PC's exposure to it is constant. On the other hand, the Internet also provides the means for maintaining your software shields at maximum usefulness with minimum effort.

Fill the Trash Some disk cleanup programs are useful investments. Windows has a built-in storage recovery feature, but add-on programs provide more options for this task. They go through temporary storage areas to weed out files that you are not likely to need again, chief among them the files that your browser captures into cache folders as you surf the Web. With a high-speed Internet link, the apparent surfing speed you gain with these quick-recall files is negligible. Also, they provide a picture (literally) of what pages you've been surfing to. You can configure your browser to limit the space dedicated to them or toss them periodically. Windows has such a facility built in that you can run from the hard drive Properties menu (see Figure 5-6). If you are running out of room on your hard drive, you may be surprised at how much of it is tied up in idle files. Some cleanup programs put these and other data into the system's trash bin instead of deleting it outright, so it's a good idea to run the cleanup before you empty the trash.

Empty the Trash Some users configure their PCs to make cute noises when the trash bin is dumped, such as buzz saws, paper crumpling, or the perennial toilet flush. If you wonder what actually happens to all of that data when you empty the system's trash bin, the quick answer is: *nothing*. That's why later versions of Windows refer to it as the *recycle* folder instead. The system's *file allocation tables* (FATs) are altered slightly to note that

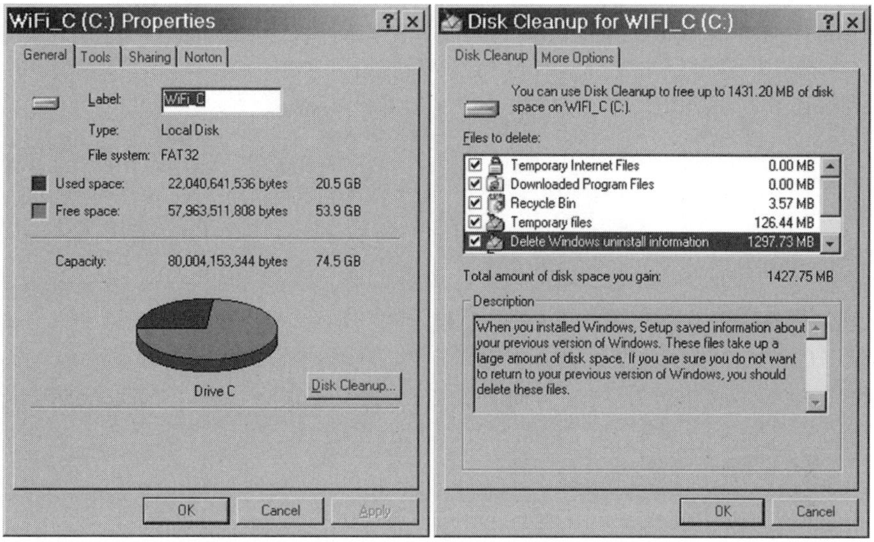

Figure 5-6
The built-in Windows program used to recover disk space

the space previously occupied by the files is now available for reuse. The data is still on your hard drive, which is why *unerase* programs can recover them, if they haven't yet been written over. In fact, some add-on utility programs auto-configure themselves to deliberately hold out the files that you think you've deleted for a specified period of time. These protected files will stay in the trash until you force them to go into the *bit bucket*. The number of protected trash files can quickly accumulate into the hundreds on an active system. They eat up space and time as well if you must scan them for errors and viruses.

Burn the Trash If you operate a home business or telecommute to a corporation from home, you will probably want to do something about these stay-behind files. Some corporations demand that employees periodically clean house, so that if their laptops are stolen or snooped, cybernetic dumpster-divers will have nothing to discover. If you work from home or carry your business data about in a laptop, it is prudent to purchase a file shredder program. In Chapter 4, we describe utilities that erase—truly erase beyond recovery—the data you discard. This information can hide in places you otherwise cannot access, such as system swap files, or file *slack space*. Until it has been written over, it can be recovered.

Check the Hard Drive(s) for Errors Windows has a built-in utility that will check the files and directories on your hard drive for consistency,

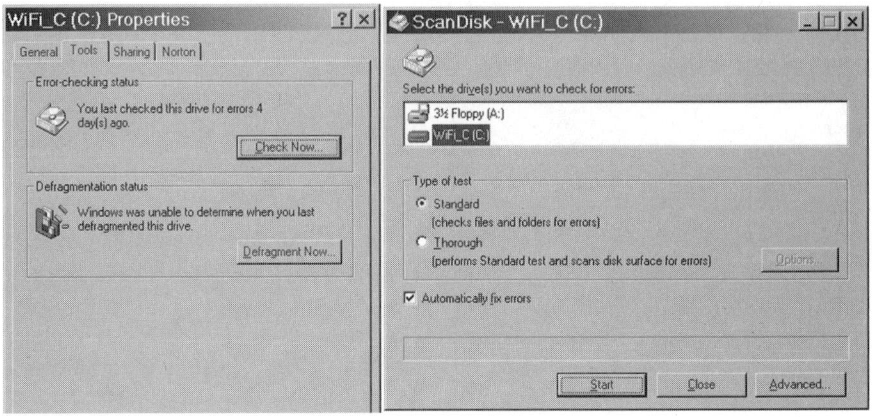

Figure 5-7 How to check the drive for errors in Windows ME

and even repair them when they are discovered (see Figure 5-7). You may have seen it engage automatically when your system rebooted following an unexpected system halt. You can start it yourself whenever you like:

1. Double-click to open the desktop icon My Computer.
2. Right-click the icon for your hard drive(s), such as C:.
3. Click Properties.
4. If you want to recover disk space, click the Disk Cleanup button. Otherwise, click the Tools tab.
5. The window that opens tells how long since the last disk error check and defragmentation. The buttons provided will start a disk error scan or a defragmentation.

At that point, you are given a choice as to how thorough you want to be as you examine the drive data. Checking the directories against the actual disk contents may take only a few moments. The program can run unattended if you check the Fix Errors Automatically box. Auto-fix is a good choice if you've decided to verify that the drive can successfully read all sectors. That process can take hours to complete. If a bad sector is found, the system will try to read it multiple times, and if successful, the data is relocated to a good sector before the bad sector is tagged so that it will not be reused. If you run this test multiple times and continue to discover new bad sectors, it is time to replace the hard drive.

Defragmenting the Hard Drive As we mentioned, when you delete a file, your system marks the disk sectors as available for use in other files.

When your system needs space to accommodate a new or growing file, it will pull sectors from this pool. Over time, files may be constructed from sectors scattered randomly about the hard drive. In practice, a file is read as you might read a long article in a newspaper that starts on page 1, resumes on page 4, then skips to page 8, and so on. Your PC's disk operating system knits them together seamlessly so that to your eyes the information in a file is contiguous.

As files become more and more *fragmented*, however, it takes the hard drive more and time to read them, as you might if you had to turn newspaper pages after every few words in order to follow a story. It takes far more time for the drive to stop reading, reposition the pickup heads, wait for the disk to come around, and then resume reading than if the heads were in one spot and just incremented slightly from time to time. The problem becomes more acute if the hard drive has to supply several programs at once. To the user, the drive progressively gets slower and slower over time.

The defragmenter starts copying every file at its beginning sector through its last and puts the copy into an empty space so that every sector in the copy is back to back with no intervening spaces. The original is replaced with its condensed copy and afterward the hard drive will read the file without having to jump around. It appears to speed up.

Add-on defragmenters from other vendors perform additional functions as well, such as optimizing disk directories, putting frequently accessed system files on the center of the hard drive where the pickup heads most frequently are, and altering programs so that they load into the PC's memory more quickly at start time. Some will also erase the empty space that results for the sake of security. In theory, your system will run reliably if you never defragment, but it doesn't hurt to do so.

Bug Patrol

Makers of antivirus software advise their users to scan for it at least once per week. Most of the programs can be set up to do this task every night, overnight, at no inconvenience. But new viruses appear almost every day, and a scan won't catch them in time unless you regularly download new definitions from the vendor's web page. These definitions are mathematical descriptions of new viruses that the scanner software uses to pick out infected software from the thousands of necessary programs on your PC. The scanner programs can be set up to refresh subscriptions automatically as well, but it's a good idea to visit the vendor's update page every so often to peruse the headlines about new bugs in town. Here's why eternal vigilance is the price of computing:

- If you get a virus, your home network can spread it quickly among your other PCs.
- A virus is much easier to block than to remove.
- If you operate a home business and give a virus to a customer, they will never forget it.
- If you take a virus from home to work, the damage to your corporation can be astronomical.
- If you have never encountered a computer virus, you are very lucky and are much overdue for one.

Other Refreshments The process of automatic periodic updates from a vendor-supplied central source works well. It can be applied to other protective software you may use. Personal firewall software relies on definitions just as antivirus software does. Parental control software, as we also described in Chapter 4, scans input from the Internet to filter out material your youngsters should not be looking at. But the list of objectionable web sites changes from day to day. If you've purchased a bundled package of these utilities from a single vendor, they can usually all be updated in one automated session.

You may employ other programs, such as Ad-Aware, to comb through your PC to weed out accumulated spyware. These web-washers also require source files, although new updates for them occur less frequently.

Security enhancements or other patches for your computer's operating system are distributed frequently in response to ever-evolving threats. Windows provides a one-click path to their product support page from the Tools menu of the Explorer browser. It makes sense to click it periodically. The manufacturers of your PC's peripheral cards may have updated drivers for them. Downloading them from their web sites may be one of the few free ways to increase your PC's performance.

Cycle out Passwords As we mentioned previously in Chapter 4, your maintenance routine should include refreshing the encryption keys used by your Wi-Fi router and workstations. How often you do so is up to you. If you operate a business from your home and you must give a password to an employee or to a customer for some reason, you cannot take it back afterward. You should change your own at the first opportunity. Likewise, it is a good idea to refresh your logins when you loan or sell a PC from your home network.

In this chapter, we discussed the mundane upkeep of your home network. If done correctly and regularly, it's boring work. In the next chapter, we'll deal with more exciting but unpleasant alternatives: what to do when the machinery refuses to cooperate. As Douglas Adams would say, *don't panic!*

Can You Hear Me Now?

Take these broken wings, and learn to fly again.

—Mr. Mister.

If the protective measures in Chapter 5, "Good Housekeeping," aren't enough, then you will need to know some basic troubleshooting techniques so you can get going again quickly. We discuss whom to call if you can't fix your network, and what to expect when you do call for help.

Basic Troubleshooting Methods

Home *wireless fidelity* (Wi-Fi) networking is so reliable that most users quickly come to take its continuous availability for granted. Wi-Fi's easy setup and installation also make it easy to forget just how complex the system really is. Inevitably, a rude reminder will come, perhaps in the form of a "cannot connect to server" error message pumped out during an email check. After several reboots and smacking the screen with your hand, you are left with the ringing question, "Now what?"

Who Ya Gonna Call?

Back in the days of a monolithic telephone company, one call to 611 was all it took to get your phone back in order. But those days have passed. Although you have the option of calling for help, you will have to do a little troubleshooting even to know whom to call. If the problem is a dead Wi-Fi interface card, for example, your *Internet service provider* (ISP) will not be sympathetic. If the problem is a defective cable modem, then calling your router manufacturer will only waste your time and theirs. Even if you get good advice from one of these agencies, the care of your network remains your job.

Unless you are willing to pay someone to come to your house (at a typical rate of $60 an hour or more), then you, the customer, will still have to do the troubleshooting. Put the stress of being cut off from the Web aside if you can, and try to consider your predicament as an interesting puzzle. Remember that however elusive it may be at the moment, a real cause exists. It will seem obvious and simple to you—*after* you find it.

Simplify the Problem

When Sherlock Holmes set out to solve a mystery, he was faced with perhaps a dozen suspects. So the task of finding the guilty party among them seemed at first to be almost impossible. Mr. Holmes applied *deductive reasoning.* He went from a general crowd of suspects to one in particular by determining which of them could *not* be guilty. By removing (or deducting) suspects from his list, he knew that whomever was left at the end, however improbable, had to be the bad guy. When you first discover that your link to the Internet is down, you are also faced with dozens of possible causes. Is it hardware or software? A weak cable crimp or sunspots? You can proceed in the same fashion as Mr. Holmes to cut the problem down to size. You do this by making a series of tests that progressively determine which links in the chain of communication are strong. Fortunately, it's very rare that two pieces of equipment fail at the same time.

Your own common sense can be a big help. If your system dies after you change it, then undo the last change to see if it restores. If so, you may have caused a resource conflict by adding hardware that's trying to use the same interrupt or *input/output* (I/O) ports as the Wi-Fi interface. It can also happen if you add new software that's claim-jumping memory space used by preexisting programs.

Dealing with Feeling The most common problem first-time sleuths have is not technical, but emotional. You will have difficulty dealing with the nagging sense of mystery. Trouble desk agents know from daily experience that every problem has a cure if you keep after it long enough, but end users lack confidence that they will ever successfully unravel the brainteaser of "might be" to the end game of "it is." You might think that the job requires a rocket scientist or someone with magical test equipment. That's why many customers call the technicians at the support desk as their first instead of their last resort. In addition to direction, they want a temporary friend to hold their hand.

Professional trouble techs have other emotional advantages. They can be coldly objective. They know, even if their callers do not, that they didn't cause the problem. So they may be concerned, but they won't be guilty. And they are perfectly willing to pursue avenues that might be embarrassing to the caller, such as "Have you installed any new software or equipment in the past few days just prior to the start of this problem?"

Sometimes service advisors share common disadvantages with customers, but for differing reasons. One of those is pressure to do something, anything, quickly. You may be hearing protests from your kids who can't play their games, chat with their pals, or even (horrors!) get their homework assignments. For his or her part, the tech is watching seconds tick by on the bottom of his or her screen while you are describing the problem. This inevitably results in the temptation to skip steps. Sometimes that hopeful gamble does pay off in the form of instant gratification. More often, it leads to a blind alley because another one of the professional's emotional tools is a disciplined approach based on scientific methods.

ISP Online Help It is true that trouble desk representatives may have some special knowledge that you lack, but if your dial-up connection is still working you can get to a lot of good information on your own. (If your dial-up won't work either, well, there you have a major clue.) Usually, advisories are posted on your ISP's web site in two forms, depending on the duration of the known problem. *Frequently asked questions* (FAQs) or white papers detail long-term problems such as bugs in software that won't be going away soon. The ISP's system status page is used to announce short-term system dilemmas that come and go, such as fiber cuts or router outages in part of their system. Often while you are on hold waiting for a tech, you will hear a recorded recitation of these bad news events. When you deduct these late-breaking causes, only a few will remain to be discussed with the professional. For example, one tech rhetorically asked us, "They were installing patches on the routers last night. I wonder if that could have something to do with it?"

Troubleshooting Tools

You already have many tools at hand, but the most important tool for weeding out problems is the one between your ears. (The fact that you purchased this book is confirmation that it is up to the job.) The second most important tool is an organized method for attacking the problem. The first step in that process is gathering information, often from your own recollection. The best time to gather is before the problem happens.

Record Keeping. As the old saying goes, knowledge is power. Detailed knowledge of how your system was built is a powerful tool for getting it back into shape. A free PC inventory program is downloadable from www.belarc.com that will tell you pages of information about your computers. Here is a list of other facts you should know about your Wi-Fi network that will help. It is more enjoyable and accurate to assemble these at your leisure than to dig them out in an emergency.

- **Know your system.** After you've completed your initial setup and all the network hardware is working well, take a few moments to write down the model and serial numbers of all the pieces. You'll need those if anything has to be shipped back for repair or if you have to locate software patches for them from a web page. Also inventory the operating system and driver versions you are using. You may be able to recite all that from memory, but someone else may have to call the problem into tech support for you.

Another good time investment is to note the appearance of the indicator lights on your equipment when it is running normally. A quick sketch will do. That way, later on you won't be puzzling over whether the enet activity light really should be on solid, blinking, or out.

- **Label the cable.** In several ways, you can build in fixability when you first install your system. Applying labels might seem to be a waste of time. Because new networks are simple at their beginning, the need for labels may not be obvious. To illustrate, your Wi-Fi router, wired hub, and modem might all use identical power plugs, even though those provide different DC or AC voltages. As you plug them in one by one at installation time, no confusion occurs. If you unplug them all for troubleshooting, you will be amazed at how exactly they resemble each other after an hour passes. If you have multiple CAT5 cables with crossover patches mixed into the "spaghetti factory," the problem multiplies. You may be able to quickly sort it out in daylight, but if you are bleary eyed after midnight, it's very easy to put the right plug in the wrong hole, thus complicating the troubleshooting process instead of simplifying it.

  If you don't know where a wire is coming from, then label it with a question mark. That's a start, because it helps to understand how much you don't know. The type of label is not as important as their presence. We prefer white electricians' tape and narrow-point permanent markers of varying colors. These homebrew tags are easier to read and less expensive than the plastic "zipper" tags you can buy at electronics stores, and one roll will supply you for years. Tape works well on flat objects too, such as transformers. Avoid masking tape. It's cheap, but it dries out and falls off.

- **Draw a map.** As with labels, a professional *local area network* (LAN) installation contract is not considered fulfilled until the documentation for it is perfect. Yours doesn't have to be perfect, but a sketch of your system, even a crude block diagram, can be a cheap and effective way to keep your investigation efficient (see Figure 6-1). Put a cloud on one end, and label it "Internet." Put a square on the other end of the paper and label that one "PC." Then draw blocks and lines representing all pieces in between and the links between the pieces, be they wired or

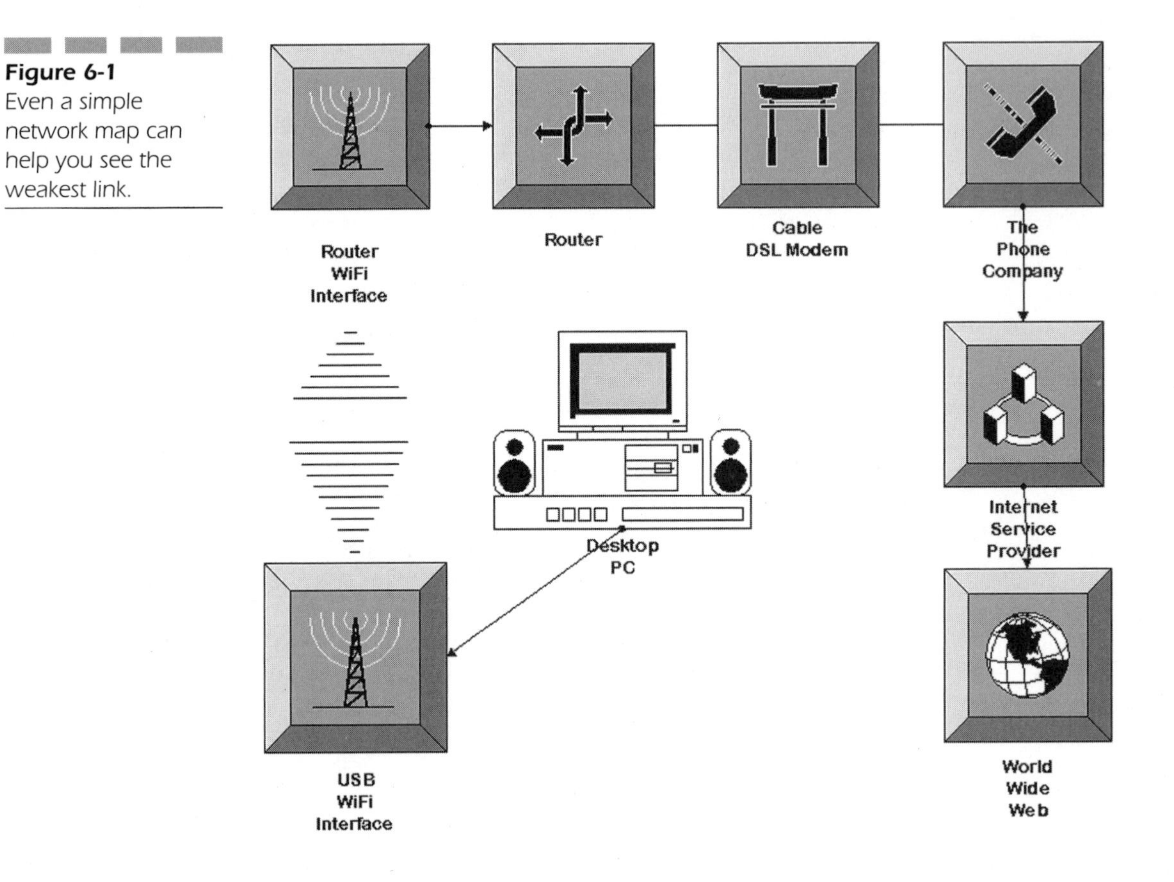

Figure 6-1 Even a simple network map can help you see the weakest link.

wireless. This simple act will greatly assist you in organizing your attack by displaying all the things that need to be checked in order. After you've checked one, you can cross it off and go on to the next. That's the simplest way to document your testing (a mandatory part of the scientific method).

- **Take notes as you go.** Writing it out often clarifies the process in your own head or suggests answers or other lines of inquiry. If you are interrupted, notes will help you for review purposes when you begin again. If you have to call for help, you can assume that you will be asked what you've done so far and the results you got. Your notes will give you confidence and save you time and duplicated effort. Details committed to paper will not fade and are particularly useful with intermittent problems, which reoccur without an immediately obvious pattern.

Quick Fixes Your own eyes are good test instruments. Check the indicator lights on your equipment to see if any that should be *on* are *out* instead, particularly power *light-emitting diodes* (LEDs). One advantage to quick fixes is that they are quick, so you won't waste much time on a few look-sees. It's surprising how often the problem turns out to be a plug or heavy power transformer that has fallen out of its AC socket. Some customers complain that trouble desk technicians are insulting their technical expertise when they say "please make sure it's plugged in." Often they discover a moment later that something *wasn't* plugged in. Techs work from scripts and ask the same dumb questions over and over again because their experience has proven to them that supposedly obvious questions are not always so dumb.

Rebooting does help sometimes. The phone company will occasionally work on your DSL line and run automatic tests on it periodically. Your ISP has to reboot the *Digital Subscriber Line Access Multiplexer* (DSLAM) on the other end of the line for troubleshooting their own system, patch installation, user redistribution, or adding new facilities. All these brief interruptions in service can leave your DSL modem permanently confused. If you decide to power reset all your Wi-Fi network equipment, perform the following steps in order:

1. Turn off the DSL or cable modem. You may have to unplug it.
2. Turn off your Wi-Fi router/access point.
3. Turn off your PCs.
4. Turn on or plug in your DSL modem or cable router.
5. Wait for all the LEDs to settle down.
6. Plug in or turn on your Wi-Fi router.
7. Wait for the router's LED display to stabilize.
8. Turn your PCs back on.

Observe the lights as the devices go through their *power-on self-tests* (POSTs). This process may take a few moments to complete. If the lights look normal afterward, see if the jam has cleared.

Another quick fix carries the idea of a cursory search a little further. Integrated circuits have become so reliable that most physical system failures are due to electromechanical breakdowns. That is, two metal pieces (moving parts, usually) that have to touch in order to pass a current are no longer touching, either due to wear or because some removable part has been dislodged. The cure for this malady is often to *push and wiggle*. Push all the connectors back in their sockets, wiggling them to make sure that they seat firmly. If these are LAN cables, watch the sync LEDs on the router and on the associated interface card. If these lights blink in time with the wiggle, then fetch another cord or try a different router port. If none of these shortcuts help, let us proceed to eliminate some suspects from the lineup.

Ping Ping stands for *Packet Internet Gopher* (or *Groper*, depending on whom you ask). It is a simple, handy utility that sends out an inquiry to a given *Internet Protocol* (IP) address and reports an acknowledgement. Every IP device will answer a ping request if it can, including yours. Much of the time, that's all you need to know, along with the IP address of your target, of course. Most likely, you began your hunt when you discovered that you could not reach some resource on the Internet. Check your map of your own system. It will show the route and links a packet would have to traverse in order to make a successful round trip to that resource. Let's use ping to test those links one by one.

The output shown in Figure 6-2 is from the DOS-based version of ping. Many *graphical user interface* (GUI) or Windows-based versions are available on the Internet for download and some of those are free. They all report the same basic statistics you see in Figure 6-2, including round-trip time, test packet size, and *time to live* (TTL). Most versions will optionally enable you to

- Change the size of your test packet.
- Adjust TTL, the interval until a timeout is declared.
- Pick the number of times the test packet is repeated.
- Send to a list of targets.
- Count the intermediate way stations (hops).

Figure 6-2 Pinging the Wi-Fi router. Can you get there from here?

```
C:\WINDOWS>ping 192.168.1.1

Pinging 192.168.1.1 with 32 bytes of data:

Reply from 192.168.1.1: bytes=32 time=27ms TTL=150
Reply from 192.168.1.1: bytes=32 time<10ms TTL=150
Reply from 192.168.1.1: bytes=32 time<10ms TTL=150
Reply from 192.168.1.1: bytes=32 time<10ms TTL=150

Ping statistics for 192.168.1.1:
    Packets: Sent = 4, Received = 4, Lost = 0 (0% loss),
Approximate round trip times in milli-seconds:
    Minimum = 0ms, Maximum =  27ms, Average =  6ms
```

- Enter the text name of your target, which is resolved into an IP address through a *domain name server* (DNS).
- Some have cute cartoons that depict the packet bouncing like a tennis ball, flying bird, or radar beam.

All we need to know is whether you can get there and back again from here. It takes so little time to run this test that you won't waste much in any case. Even so, we like to cut the problem in half with each test if we can. In this example, we've picked our Wi-Fi router as the midpoint, and as you can see, it does answer. This means that all the links between the originating PC and the router are good, and you can cross them off your list of suspects for now. If you did *not* get an answer, put question marks on all the links, and test them one by one. Remember that PC firewall software is one link, so disable it while testing. You are allowed to test your own PC's address too incidentally. The test just loops internally in that case, but if it does not, then you know that your PC has a problem, possibly its network configuration.

Parts Substitution Another tactic to try if you get a null result is to replace a suspect link in the communications chain. If you can't reach your favorite web page, for example, try another web page before you assume the problem is local. Since you have a home network, you could also repeat your series of pings using another PC on that network. You may not have a spare router in the closet, but you can bypass the one you have by connecting a single PC directly to your cable/DSL modem as you did when you first set your network up. In this direct-connect mode, you can try pinging the cable or DSL modem. Check the documentation for it. Most were given nonroutable addresses at the factory (such as 192.168.1.1) just like your router/access point.

In the same way, you may not have a spare Wi-Fi interface card, but you can bypass the airwaves and temporarily hardwire one of your PCs to the router's RJ-45 ports if none are already connected that way. No power indicator? Try another AC outlet. How would you go about replacing your entire network? Log into the Internet by using your dial-up modem. If that does not work, either your login has been disabled (Eureka!) or the ISP has been hit with an earthquake. If your local network tests clean, you can also use the dial-up connection to ping your router from the outside in using its *wide area network* (WAN) IP address as the target.

The important thing to remember in this process of progressively testing, swapping, and bypassing is that you must run your tests one at a time and take notes after every test. If you change more than one condition, you won't know which change made the difference when the pings start coming back to you. The first rule of a good surgeon is *do no harm*. Be patient and do not blindly tear into your system for fear that you might make it worse.

When you've isolated the trouble spot to one particular piece of hardware, you may wonder how you go about fixing it. If the hardware involved is anything more complicated than a piece of wire, don't bother. You will need special test equipment to find a bad chip and no user-changeable components are inside. Most of the time, it is under warranty from the manufacturer or the ISP. Most of the time, if the box is really dead and you have no alternative, it's cheaper to toss it and buy another.

Windows Command-Line Utilities *Tracert,* the trace route command, is like ping in several ways. It is invoked from a DOS window. Many GUI versions of it are available for download for free. Given an IP target, the program sends out an inquiry, but unlike ping, tracert will put a stopwatch on all the intermediate steps on the pathway to the target (see Figure 6-3). When you try it, you may be surprised to see that whenever you surf your favorite web page, your traffic must pass through a dozen routers or more to get there and back, and the route changes over time.

Tracert is most useful in troubleshooting connections outside of your own home network because that's where most of the relay points will be. Game players could use this command to determine why their connection seems so slow. They can pinpoint the slow link and also find out if their ISP routes

```
C:\WINDOWS>tracert mcgraw-hill.com

Tracing route to mcgraw-hill.com [198.45.19.151]
over a maximum of 30 hops:

  1    13 ms    28 ms    14 ms  hsa001.pool024.at101.earthlink.net
  2    14 ms    28 ms    27 ms  216.249.64.33
  3    14 ms    27 ms    14 ms  207.217.2.129
  4    27 ms    14 ms    14 ms  209.165.101.2
  5    13 ms    28 ms    13 ms  POS6-1.hsipaccess1.LosAngeles1.Level3.net
  6    27 ms    27 ms    14 ms  ge-6-0-1.mp2.LosAngeles1.level3.net
  7    28 ms    27 ms    41 ms  pos9-0.core2.LosAngeles1.Level3.net
  8    27 ms    27 ms    28 ms  POS4-2.BR1.LAX9.ALTER.NET [204.255.169.105]
  9    27 ms    28 ms    27 ms  0.so-0-1-0.XL1.LAX9.ALTER.NET [152.63.113.10]
 10    41 ms    42 ms    41 ms  0.so-1-0-0.TL1.LAX9.ALTER.NET [152.63.115.142]
 11    96 ms   110 ms   110 ms  0.so-2-2-0.TL1.NYC9.ALTER.NET [152.63.1.118]
 12   138 ms   123 ms   124 ms  0.so-3-0-0.XL1.NYC1.ALTER.NET [152.63.27.29]
 13   110 ms    96 ms    96 ms  0.so-0-0-0.XR1.NYC1.ALTER.NET [152.63.19.85]
 14    96 ms    96 ms    96 ms  507.ATM8-0-0.GW3.NYC2.ALTER.NET [152.63.20.33]
 15    97 ms   109 ms    97 ms  mgh-t3-gw.customer.ALTER.NET [157.130.18.70]
 16   109 ms     *       96 ms  gw2.mcgraw-hill.com [198.45.19.20]
 17    96 ms    96 ms    96 ms  198.45.19.151

Trace complete.
```

Figure 6-3 Tracing the route. How long does it take to get there from here?

their traffic to the far side of the continent before putting it on an Internet backbone. Let's look at some other command-line utilities.

NET VIEW lists the computers on your network with *Network Basic Input/Output System* (NetBIOS) support. If you run this command and see a computer that's not supposed to be there, it's time to change your *Wired Equivalent Privacy* (WEP) parameters. This will be the first place that a Wi-Fi squatter shows up. It will also show you if you are inadvertently connecting onto someone else's Wi-Fi LAN.

NET VIEW *name* will list all the available shares for the computer name you specify. A hacker who has just found his way onto your Wi-Fi network might use this command to see what he can get into, assuming that you've shared the offerings without passwords. Using Net View with the "/?" or help parameter will give you other run options. For the following programs, you would use "-?" to get the same information.

If you want to see which IP ports your PC has left open and which programs opened them, then enter the *NETSTAT* (NetBIOS statistics) command with the -a parameter. You can use this to look for an active Trojan horse program or to learn which ports must be opened in your firewall to accommodate a game. The -e (for Ethernet) parameter gives you a running total of your packets in and out. The -n (for numeric) parameter displays the IP addresses and port numbers that your PC is connected to. If you use a Wi-Fi router, most of these are going to be nonrouted addresses assigned to local computers by the router. Use this to find out if your PC is automatically connecting to another's file system.

Your PC connects to the others on your Wi-Fi net by their NetBIOS names or their IP addresses, depending on the application you are using. Your PC keeps a table that maps addresses to names and keeps track of who is connected. The *NBTSTAT* (NetBIOS over TCP/IP Statistics) command dumps that information in various ways. The -s parameter, for example, gives you a list of your current network connections.

Windows IP Configuration (WinIPcfg) *Windows IP configuration* (WinIPcfg) can be called from a DOS window, but it gives a GUI display. You may want to create a shortcut for it on your desktop, which will save you steps. It quickly reports a lot of useful information about your network connection, as shown in Figure 6-4. This command is the simplest way to determine the *Media Access Control* (MAC) (adapter) address of your *network interface card* (NIC) before you enter it into the router's access list. Also shown is the address of the Wi-Fi router or gateway.

In addition to reporting information, the command can also force your Wi-Fi or wired NIC to release (forget) its IP address and fetch another one

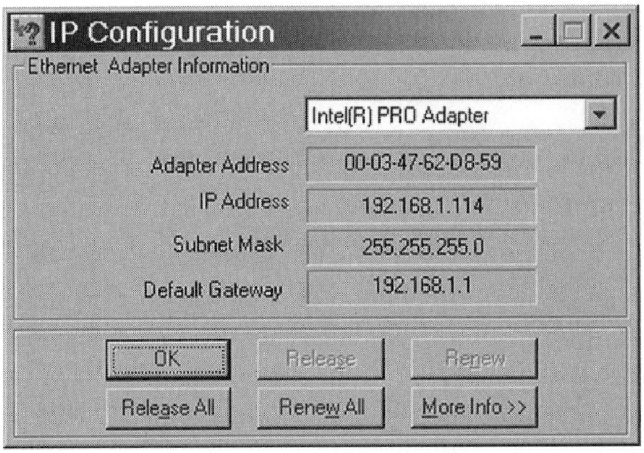

Figure 6-4 WinIPcfg tells all about your network connection.

(renew) while you watch. To do so, it must make a *Dynamic Host Configuration Protocol* (DHCP) request to your Wi-Fi router, as it does during the PC's boot sequence. A successful release/renew command is a simple way to verify that

- The NIC is working.
- The PC is attached to the net.
- The PC has the correct address for the router.
- The net connects to your router.
- The router is working, at least in your direction.

Wi-Fi Router Status Display

To learn about whether the router is working in the other direction or out toward the Internet, you can check the router's status screen. Wi-Fi access points and routers can tell a lot about themselves. So if they will talk to your web browser at all, they are a useful diagnostic aid (see Figure 6-5). If you cannot connect to them from any of your PCs, wired and wireless, that fact is also a loud diagnostic message. If a power reset doesn't help, you may have to totally reset the unit. This is often done by pressing a recessed switch while cycling the power, but check the manual for your particular model. The action restores all the factory defaults, including passwords, so if the router starts working afterward, any configuration work that you've done will have to be repeated from the start. The reset switch is usually protected to prevent this from happening unintentionally. You did remember to write down your router configurations, didn't you?

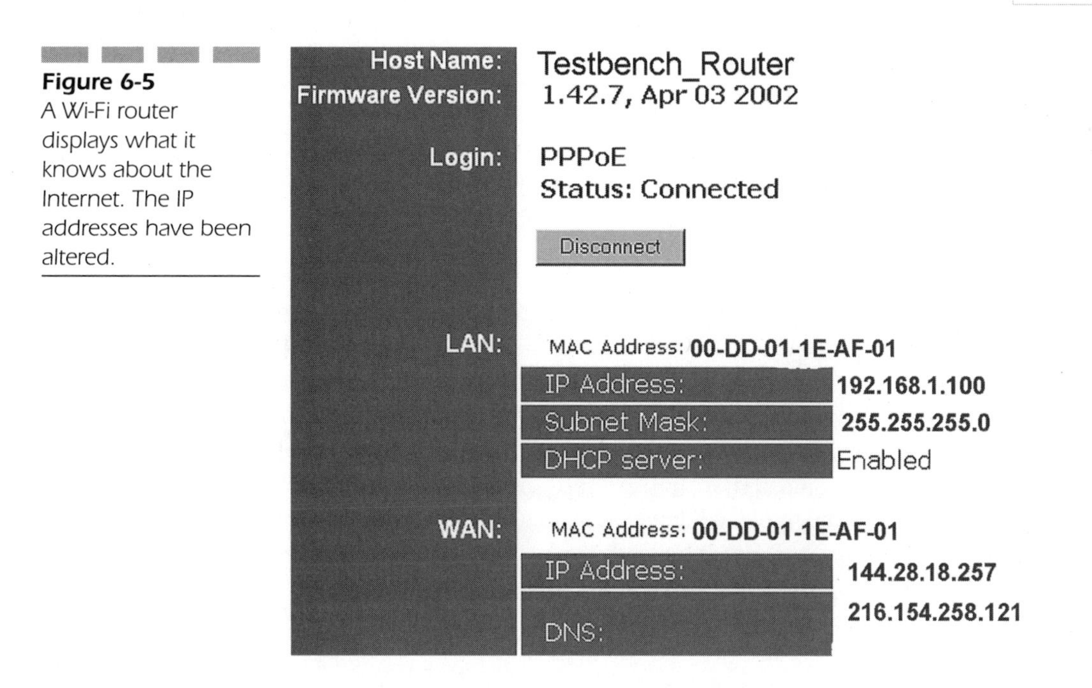

Figure 6-5 A Wi-Fi router displays what it knows about the Internet. The IP addresses have been altered.

In the previous display, the most important information is that the Wi-Fi router says it is connected and logged in to the ISP properly. You can force it to disconnect and reconnect, as with the winIPcfg command. If successful again, you've proven that your router is connected to your DSL or cable modem, the modem is working, the outside line is working, and the ISP is answering by logging you in and providing a new or renewed IP address.

The IP addresses are also important. The local, nonroutable one that usually starts with 192.168.*x.x* is what you would ping from a local PC to test connectivity to the router from the inside. Your router's WAN IP address is what you would use to ping your router from a dial-up connection or a friend's PC in order to verify that the line out to the Internet is working.

If you call the Wi-Fi router manufacturer's help line, they may ask for the *firmware* (semipermanent software) version of your router. The main status screen is where you will find it. If you are instructed to upgrade the firmware to the latest version, the process is simple:

1. Get the upgrade instructions specific to your equipment. They vary.
2. Turn off your PCs except for one that is hardwired to your router.
3. Use your browser to connect to the manufacturer's support web page.
4. Look over the available downloads. Choose the proper file for your router.

5. Copy the file across the Internet into a temporary directory on your own PC.
6. Execute (run) the file.
7. You may have to power-reset the router afterward.
8. Check the status page to verify that the version number has incremented.

A blinking LED on the router usually indicates upgrade progress. Never interrupt the process before it is completed. You might wind up having to send the router in for repair if you do.

Some DSL and cable modems memorize the MAC address or the unique hardware identifier presented to them by your network when they are first turned on. ISPs can interrogate them via the Telnet program. It doesn't happen often, but modems may stop working when you replace a single PC with a router, because it has a different MAC address. That's why routers give you the option of changing their WAN MAC address to a clone or match whatever used to be there. MAC addresses aren't as unique as they used to be.

Dealing with the Trouble Desk

The service may be known as technical support, problem resolution, customer assistance, or other names, but the common industry designation is the *trouble desk*. Let's assume that you think you know what the problem is or you've run out of things to test. Your hunt is at a reasonable point to phone for help. Generally speaking, the smaller the organization involved, the more responsive it is likely to be. So if you are confident that the problem is a broken NIC or wireless router, you'd call that manufacturer's help line. Your first point of contact for anything else, including a suspected phone line problem, is your ISP.

Manufacturers and ISPs alike advertise technical advice, most of them on a 24-7 basis. But other than advertising the service as a value-add, call centers are an operating cost, not a profit-builder. Providers try to cap those costs in whatever ways they can. The results can be frustrating, even infuriating to a stricken customer who is often anxious and upset to begin with. ISPs do not advertise the average delay between call placement and the moment a technician actually says hello. (We once had to wait 45 minutes to get a password changed.) The wait time can vary widely depending on the center's call load and is often longer than the time it takes to diagnose a problem. They are busiest on Mondays and Fridays.

Call center mangers know from moment to moment how many users are waiting for help and how long they've been holding. They get detailed reports listing how many called at what hour of what day, how long they had to wait, and how many chose to give up while waiting (known as *abandons*), plus a breakdown of the types of trouble reported. Many call centers routinely assign *secret shoppers* who pose as customers to sample their *quality of service* (QoS). It's true that their call load will sometimes spike due to a major outage, but usually the time you spend on hold is the result of their management's premeditated cost/benefit analysis. Do not presume to place a trouble call as the last thing you do before leaving the house for work or a movie, because you may have to hang up and depart before your call is answered.

Many of the tips to follow can be applied when you call for help about a broken washing machine or missing paycheck. The troubling issue may be different, but the basic principles of the exchange are universal.

Dialing for Answers

You can take certain measures in advance to speed the process:

- **Know yourself.** Trouble desk advisors must ask identifying information, such as your phone or social security number, part of your email password, or your mother's maiden name. Otherwise, an impersonator might get them to change a password or otherwise compromise your security. If your ISP is not the same as your DSL provider, you will also need to know the DSL provider's exact name; "the phone company" is not good enough for a tech who may be on the other side of the continent.
- **Do your homework.** As we mentioned previously, your provider probably has tutorials, a FAQ, and troubleshooting tips on a web page. Checking this advice can solve many common problems and save you time. Your equipment inventory and the notes you took during your initial troubleshooting should be close to the phone. Also keep your notepad handy, because you will continue to need it.
- **Hands free.** If you have the option, use a speakerphone. A cordless phone with a headset is ideal. It will enable you to perform tasks as directed by the customer agent and talk without interruptions. You can also continue to work while on hold, which helps to keep you from feeling as though you've been strapped into a timeout chair. You do have one consolation: Because you have dialed an 800 number, you are not being charged for the privilege of holding.

Navigating the Maze

The routine varies, but all service providers have depersonalized their support, just as the banks and utilities have. An *Interactive Voice Response Unit* (IVRU) greets everyone who calls the main number. It also collects statistics regarding how many call, when they call, and what selections the callers make. It may also collect caller phone numbers, using caller ID. The larger the organization, the more complex their IVRU is going to be. You will hear it as a recorded voice, usually female, that assures you that your call is important and then asks you to listen carefully to the following menu choices. A good reason to listen carefully is that if you push the wrong button, you will *not* go to the head of the line when you are transferred to the right department. You will resume the elevator music from the beginning. So, as the menu choices are recited to you, write down the one you want on your clipboard. Typical initial menu choices are as follows:

- Sales of new accounts, new services, service upgrades
- Buying software; time; or baseball caps, coffee cups, or T-shirts bearing the company logo
- Business users
- The extension of an individual
- Tech support
- Billing disputes
- A number that replays the menu

By the time the recording comes to the end of the list, you may have forgotten what the choices were. This is easy to do if you are fatigued after working late on your broken network. Sometimes you can get connected to an operator by punching 0 or stubbornly not choosing anything, but don't count on it. When you push the number for technical help, you won't get technical help. Instead, you'll hear another IVRU menu asking you to choose between

- Password changes
- A number that kicks you back to the first menu
- Technical help from a robot, which is an audio readout of the walkthroughs and advice available in text on the web page
- A list of dial-up access numbers by city

Choosing to speak to a human technician will instead fetch yet another IVRU menu, asking if your problem has to do with

- Your web page
- Your DSL connection
- Your satellite connection
- Your wireless or home network (They are referring to network equipment they sold you. If you didn't buy any from them, don't push this button.)
- Your dial-up modem

Let's assume you've selected a DSL connection. You may be asked to choose from a variety of operating systems, what part of the country you are calling from, whether you are setting up your system for the first time, or if you have actually used it before it went dead. The IVRU will eventually decide where to route your call. When it does, you may be warned that your call may be monitored. If a regional or system-wide failure is in progress, you will hear another announcement assuring you that the engineers already know about it and are working on it, so if that's why you called, you can hang up. If this hasn't headed you off, you can expect to hear the following announcements, repeated at intervals:

- A mechanical voice telling you that all the techs are busy, so you will have to wait for *X* minutes.
- A recording reminding you that you can always ask your question by email, consulting the web page, or reading the instructions. (Reading the manual is a reasonable request.)
- A recording asking you to please unplug your equipment and try it after plugging it back in. If that works, hang up and have a *super* day! (Often, it works.)
- A recording asking you to please stay on the line.
- Generic music. This can be Muzak: great show tunes of the fifties, themes from Hollywood movies, or midi tunes from a Casio tone player.

You will have to keep an ear open even if you are working at something else. If a customer support agent announces himself or herself and gets no immediate reply, they will assume that you've hung up, and so will they. They are on the clock throughout their shift and cannot afford to wait.

Trouble Desk Etiquette

Now that you've negotiated the maze and connected with someone who can listen to you, here are some tips regarding how to deal efficiently with the trouble desk engineers:

- Foremost, keep in mind that none of this is their fault. They did not write the call center practices and scripts. They did not choose the so-called music you had to listen to. They understand that waiting is humiliating for most people and particularly aggravating for customers who've had to call back several times. Given a choice, the techs would answer the phone on the first ring and talk all day, but unfortunately, everything they do is under the baleful gaze of a computer. You can be firm and persistent, but remain civil. Threats or sarcasm (especially obscene insults) are definitely counterproductive.
- Ask them to repeat his or her name, even if it's only a first name, and write it down. You can try to get his or her personal extension number too. You may be handed off to other techs multiple times, which means multiple names, and that's why your notepad is handy. Usually, hand-offs are a good thing because you'll be talking to someone with more experience. It can also mean that you really are getting the runaround.
- Ask for a ticket, case, or confirmation number or whatever they call it, and write that down. If you are calling back, give them the original ticket number(s) so they can pull up your history. Most techs like to open up new case numbers because it makes them look busier and also more successful at closing cases. If you collect half a dozen numbers, be sure to recite them all when you call. They are particularly useful if you are explaining yourself to a supervisor.
- The technician will ask you your name and some identification, and pull up your folder. This may contain obsolete information about you or your hardware, so you may have to correct it. You may be surprised at all that's in there.
- The fact that customer technicians must meet a quota guarantees that they will be under stress too, and anyone can have a bad day at the office. If you get one who's impatient or disrespectful, don't get mad or even—get the manager. The same rule applies if, after they've had a fair chance, you feel that they just can't help. Service reps will never admit that they don't know what you are talking about. They may not want to ask for help out of pride or because they've been discouraged from bothering the boss. But once you ask, the decision is taken from

them and they have to stick their hand up. No hard feelings. When you get to a supervisor, be calm and don't exaggerate.

- If all else fails, you can always write the CEO. Really! Chances are that the big boss will never actually see your angry-gram, but the note will be passed back down the food chain to the trouble desk manager, who will have to pay attention and report progress back up the ladder. All this takes a while.

  You can also write the *Public Utilities Commission* (PUC) or, as it is called in some states, the Public Service Commission. ISPs are not regulated in the fashion that telephone companies are. Internet providers cannot be fined, but if the state PUC gets many similar complaints, they are allowed to ask questions. Their web pages will instruct you how to file a complaint, and writing will make you feel better. You can complain to the FCC too at www.fcc.gov. The Better Business Bureau also has a web page, and they usually get respect from ISPs.

- Lastly, know when to quit. Sometimes the only cure for a chronically bad ISP is a new ISP. This is the point at which your detailed notes and email copies come in. Theoretically, the provider broke the contract by failing to provide a remedy, but you will have to prove that in order to have any hope of avoiding early disconnect penalties.

A Case History

Most of the time, DSL works flawlessly. When it fails, the cause is often isolated and repaired quickly. If a first-line tech support representative is stumped, then they must refer a trouble ticket to a second-line engineer who has more experience or other resources available. He reads the call history and checks what he can, such as router settings. If he finds the problem, the ticket is closed. Someone else from the ISP is supposed to follow up by calling the customer to confirm that the problem really has gone away. If it has not, then a third-line super-expert is summoned, and so on. But sometimes the problem returns, no one calls you back, and the cure doesn't go by the book. Here's a real-life customer log:

Monday:

Email says "connection to server lost." Can't reach any web pages either, but every PC in the house can ping everything else, wired or otherwise. Access came back on its own after an hour. A check of the Wi-Fi router's status page shows a new WAN IP. Kept looking, but nothing to find.

Tuesday, Wednesday, Thursday:

Up and down, mostly up, but with no obvious pattern. Pushed and wiggled on all the connections, found two homemade flaky LAN drops, due to stranded-wire RJ-45 plugs being used on solid-wire lines. Replaced them. Found a flaky port on the hub, which I taped over and stopped using. But connections going from the DSL modem to the outside were solid; modem sync LED solid. When system is down or up, could drop and renew my DHCP IP.

Friday:

Called tech support in the late evening, and got through in 10 minutes. Tech said they've been installing patches to the routers, but he doesn't know if my town was involved. He pinged the router at the *central office* (CO). I can ping my PCs and my router in any combination, but can't see anything on the outside. Multiple power-on resets are no help.

He asked if I had a router, and I confessed. He said he has one too on his cable modem, but I would have to disconnect mine to continue. I did so, substituting a laptop. Configured the laptop with a temporary dummy address, but still could not ping the DSL modem. He thinks the modem is dead. Then I discovered that the personal firewall software on the laptop was interfering. Turned it off and could ping the DSL modem.

The tech said try logging in. No *Point-to-Point Protocol over Ethernet* (PPPoE) software on the laptop. Tech said okay, log in using a phone modem, download the software from their web site, configure it, test it, and call back.

Saturday:

Downloaded WinPoet (their brand of PPPoE software) onto the laptop. It worked! Then I discovered that everything worked. It had come back on its own again. Reinstalled the wireless router. Reinstalled the old telephone modems on the house PCs too, in case this siege continues.

Sunday:

No Internet. No WinPoet connect. Called Tech Support again, gave them the original ticket number. The new tech wanted a second ticket number and wrote that down. Recounted everything to the new tech while he read the others' notes. He thinks something on my phone line in the house is making the DSL signal fade, although the sync LED has been steady. I unplugged phones from the wall jacks, leaving one to talk on, but no fix.

Called back, requested the same tech, and eventually got him. He said the cause may be flaky house wiring and instructed me to plug the DSL modem into the test jack at the point where the line enters the house. There isn't any test jack. He gave me his desk number and email address so I could report what I found after I made a test jack.

Called the phone company on my own and asked what it would take for them to check my inside phone wiring. (They ran an automated test on the line from their end, and it looked good.) Answer: If they visit the house, check, and find nothing, I get billed. It's my gamble. I said thanks, I will check on my own.

Took all the terminal block connections apart, sanded them clean, attached a phone jack, got an extension cord, and plugged the DSL modem into the line by itself (see Figure 6-6). Sync LED okay, but still no WinPoet connect. Tech's email address was incorrect, so left voice mail.

Monday:

WAN down in the evening, back up again in the morning with a new IP assigned. Waited on hold 30 minutes this time. I suggested they call the phone company to test CO equipment. Tech said he couldn't until we check the DSL modem. He asked for the model number. I looked for it, and then noticed that the power LED was out. Yes, the transformer is firmly plugged into the electrical outlet. Must be the modem, dead at last!

He transferred me to another tech (only five minutes on hold this time). This technician issued a *Returned Material Authorization* (RMA) for a new modem under a third trouble ticket. The new one will arrive by UPS in four days. While reassembling the network, I discovered that I'd dislodged the power connector when turning the modem over to find the model and serial numbers.

Tuesday:

Phone with tech support re: new DSL modem arrived (late) was installed OK, even though line has been up for two days. Did they

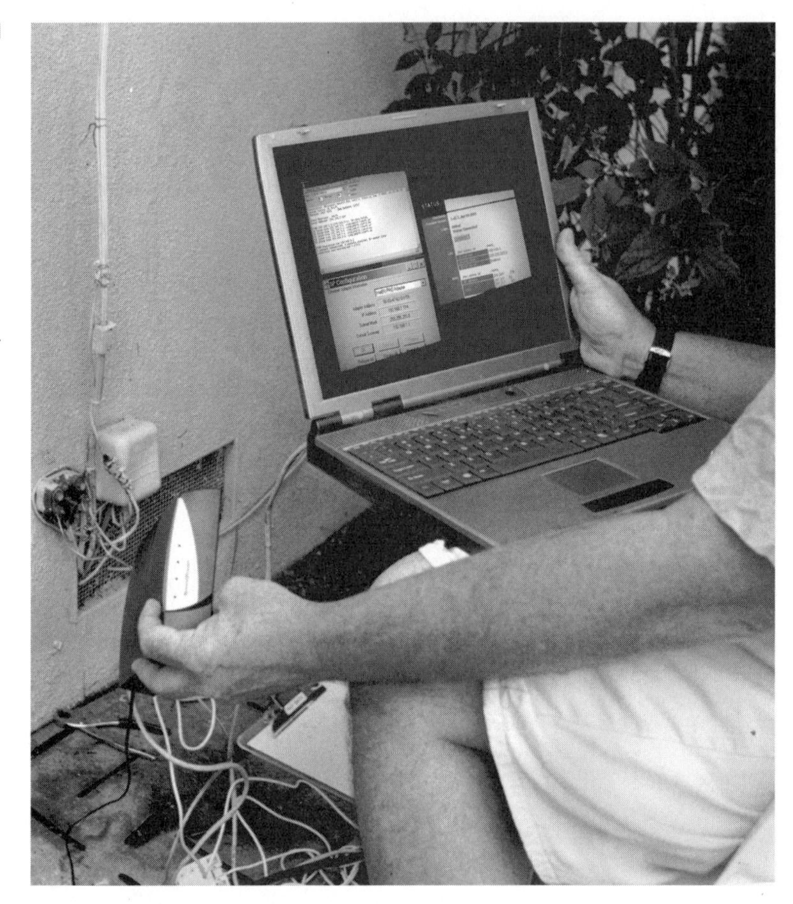

Figure 6-6 The customer disconnects his house phones entirely and sets up his DSL modem directly to the incoming phone line. Photo by Daly Road Graphics.

fix the line? Answer: No, not according to the case logs. Also, no prepaid return ship label was included for the return as promised. Tech said they don't handle that, transferred me to another department. More music. Another clerk said that they don't handle that either, wanted to transfer me back to tech support. I protested, saying that I wouldn't go unless they get new Broadway show tunes because I've memorized them all. He relented, promised to FedEx another prepaid shipping label.

Thursday:

Called tech support, gave them the four ticket numbers and the RMA number, still asked where's the shipping label. They said they don't handle that and transferred me to sales. Explained it all again to a sales rep who promised to send a label, this time by UPS. She gave me a "confirmation number," which I added to the others.

Monday:

Label arrived, modem shipped out, WAN has been solid. Cause and cure were never discovered. (Possible sunspots.)

Common Wi-Fi Problems and Cures

At this point we will assume that you've traced your Internet connection problem to your wireless PCs and that the hardwired ones seem to be okay. You can continue the process of elimination further by determining whether only one of your Wi-Fi PCs is mute or all of them. It is unlikely that more than one would break at the same moment unless the problem is really something common to all of them, most likely the access point/router. With Wi-Fi terminals you have the option of bypassing the access point for test purposes. When the interfaces are configured to run in ad hoc mode, they will attempt to interconnect with all the other terminals in range. If all the PCs can see all the other PCs, then you've confirmed the router as the sore point. If one PC is not present or accounted for, then concentrate your search there.

Unlike a LAN cable, the radio signal that connects your terminals cannot be examined for breaks or damage, but the Wi-Fi interfaces have indicator LEDs that will tell you if networking is turned on and the card is seeing a signal. The manufacturer's configuration program is more verbose, because it will report the signal's strength and all the other details of the interface's setup. One way to reduce doubt regarding signal strength is to move the PCs closer together, which is particularly easy to do with a laptop.

If you suspect bad hardware, it may be easy for you to confirm it by swapping the suspect interface onto another PC, particularly if it is designed for easy relocation as Cardbus and *Universal Serial Bus* (USB) interfaces are. If the fault moves with the card, the fault is the card. If the fault stays with the PC regardless of the interface, the fault is the PC, and most likely the Wi-Fi configuration.

Configuration Does Count

If the interface seems to be working at all, the problem may be imperfect software configuration. A Wi-Fi network does not tolerate nonconformists, particularly if you've properly engaged the security measures that Wi-Fi provides. These work by deliberately excluding any stations using the wrong *service set identifier* (SSID), encryption keys, or radio channel. Write down the values in use by the terminals that do work. (If you wind up calling tech support, they will ask you to read off those values.) Carefully check the PC that does not work to see what, if anything, is different about its setup screens. If you do find something odd, change it to match the working ones, reboot, and test again. If all your configurations are different, you might as well restore them all to factory defaults and resume from the beginning.

Common DSL Problems and Cures

If all the nodes on your home network can see each other but none can see the Internet, or if you cannot ping your router from a dial-up PC or get a WAN light to show on your DSL modem, then the problem may be on the incoming telephone line. This is not always something that the telephone company can find or fix. We've discussed how some things can soak up or diminish the high-frequency signal that carries data to your house. These include too many phones on the line or defective microfilters. But a strong signal can be covered by extraneous noise on the line as well, slowing your DSL modem to a crawl. It can originate from a distance, as with radio stations, or locally, from

- Noisy telephones
- Cordless phones
- Home alarm systems that dial out on the DSL phone line
- Phone lines close to fluorescent or halogen lights
- Noisy power lines (see Chapter 5's "Power Corrupts" section)

You can determine if the source is local by disconnecting phones or turning off lights. But some DSL line noise comes from the telephone company itself. Some examples are high-speed data services such as T1 or ISDN lines. If these share a conduit or other cable run with your DSL line, they can leak noise into it by a phenomenon known as *inductive coupling*. These sources are known as *disturbers*. One symptom is that your DSL modem slows predictably at a particular time of day. If you ask the phone company to check for noise, you should voice your suspicions and ask that they test at that time.

Residual Line Hardware

Sometimes the phone company will install a special device at your house that remotely tests your telephone line for them. It's called a *maintenance test unit* (MTU). It should have been removed when your DSL line went in, but if they missed it, it can seriously interfere with the DSL data signal.

If the modem's DSL, Sync, or WAN LED goes dark when someone picks up the phone, or overnight, the cause might be a *Digital Added Main Line Multiplexer* (DAML). They funnel multiple customers through a single phone line. They block the DSL signal or interfere with modems when the CO runs automated testing on them.

Enforced Idleness

Some ISPs will shut your DSL connection down during periods of prolonged inactivity. The green sync LED will be on, but in effect, nobody will be home. If your connection is automatically idle in this way, your router is supposed to instantly log you in again when the connection is needed. In practice, some DSL modems go to sleep in the interim. Kicking users off was appropriate in the days of dial-up modems when "campers" would sometimes hang onto incoming dial-up modems for days on end, whether they were using them all the time or not. DSL connections, however, were designed to be always on.

Most Wi-Fi routers have a setup option that periodically puts traffic on the line for this reason. Other PC-based programs such as PONG achieve the same end. System clock utilities that automatically check Internet time-servers can serve the same function, as will email programs that check for new messages every few minutes. If your modem continually needs a power on reset in order to wake up, or your PC needs a reboot for the same reason, then consider asking your ISP for a newer model.

Common Cable Broadband Problems

As with DSL modems, instances may occur when your local network is adequate for data transfers to local targets, but seems cut off from the World Wide Web. Check out these possible causes before you call the cable company.

Line Splitters

Some cable companies will increase their monthly charge for extra outlets in a customer residence. This prompts some customers to buy extension cables and splitters so they can do it themselves. If the customer is using these lines for TV only, then usually no harm is done, but data connections are bidirectional and cheap splitters only work for incoming signals. Your cable modem must be the first item on the incoming line. Putting a splitter ahead of it will either diminish the signal too much or block the upstream half entirely.

In the same way, if your cable modem was working fine until you removed a cable company splitter, the problem may be a signal that is distorted because it is now undiluted and too strong. Cable modems are very sensitive to signal levels. You may need a visit from one of their technicians, who will install a two-way *attenuator*.

Defective Cables

Coaxial cables are actually tuned *radio frequency* (RF) circuits. They can cease to be efficient signal conduits if they are cut, tied in a knot, bent at sharp angles, slammed in a door, or full of water. Cable modems, as noted, are more likely to be affected than your TV screen. If your Internet service is slow or intermittent during and after rain, then you may need to have the run from your house to their backbone replaced. Fortunately, most firms will do this for free, after they've determined that the cable run is truly inadequate. Proving this is easy to do by connecting a signal injector on one end and a signal strength meter on the other.

If your cable modem works well when a PC is attached but fails when you connect a Wi-Fi router or access point to it, the problem may be neither of them. Some cable modems memorize the MAC address of the PC that's connected to them and clam up if something else takes its place. For this reason, many routers give you the option of cloning or imitating a different device's MAC address. You may also cure it with a power-on reset to the modem.

Unsolid Connectors

The general troubleshooting method we described in the first part of this chapter should have flushed out a defective cable, but perhaps you doubted that the factory-built, two-foot patch cable between your router and the modem could have anything wrong with it. Patch cables are actually more likely than long runs to be defective, because they are moved around more often. In any case, take a moment to look at the link LEDs on the router port and on the cable modem. Wiggle the patch cord. If the lights blink or you are doubtful for any other reason, then dump the cord.

Troubleshooting Shoot-'em-Ups

Your network may seem to work from the tabletop to the global Web, but it may fail with some particular applications. One example is interactive Internet games. You might not think so, but to a veteran gamer an Internet outage ranks alongside the *Titanic* as a major disaster. We provide specific configuration examples in Chapter 7, "Wi-Fi for Fun and Profit."

Gamers don't really care to know about details like MTUs, *demilitarized zones* (DMZs), port forwarding, and the like, but the state of multiplayer gaming art is not plug and play. Internet computer games have special requirements, and so a few special fixes for them are listed here.

Most games communicate using *User Datagram Protocol* (UDP) packets. The original intent was to get quick reactions between players, because these packets don't require an acknowledgment or retransmission if they don't get through. It's not a problem on a local LAN, but crossing a local router with *Network Address Translation* (NAT) translation onto the Internet and then crossing dozens more routers to a game server invites lost packets.

Many, many multiplayer Internet games are on the market. Their particular IP requirements are not standardized. However, Internet game players are among the early adopters of cable/DSL in the home. They are a big segment of the gaming market. Vendors understand their needs and are eager to meet them. Vendor web sites have specific technical advice, and gaming web site forums are another source of router configuration tips.

The DMZ

Some games, such as those by Microsoft, open lots of ports. Some of these may be blocked by your ISP, and more will be blocked by your own router unless you unblock them (see Chapter 7's "Special Gaming Considerations"). If you have lots of PCs, you may not have enough options to go around. This leaves you with the dangerous cure-all of putting your PC in the DMZ, thus removing the isolation and the protection your router provides. Most Wi-Fi routers will only enable you to do this with one PC. If you are playing host for someone on the outside, your host PC should be the open one.

Game Servers

Many Internet multiplayer games use servers to exchange information about the competitors. Some web sites, like GameSpy, exist to provide them. They allow the same virtual playing field to be displayed to dozens of competitors. When these games are started, they usually launch many pings to locate available servers on the net. This can be a problem, because some network firewalls mistakenly perceive it as a hacker attack and respond by shutting down traffic from your WAN IP. Also, routers do have a limit on the number of connections they can maintain. A wide-open server search may go beyond that limit.

The cure is to limit the number of servers you look for at one pass, which is a configuration parameter you can adjust when the game or server site software is installed. You can do the searching yourself and highlight only a few of those available. You can put the PC in the DMZ so your router won't care anymore.

Some Internet games require your WAN IP address. As mentioned previously, you can find that on your router's status page, or you can surf to www.whatismyip.com, which will report it to you.

Chat

The ICQ Send File option can work with your Wi-Fi routers, but you must first fiddle with the program parameters:

1. From the ICQ menu, click Preferences.
2. Click the Connections tab.
3. Check "I am behind a firewall or proxy."
4. In the firewall setting, set the timeout value to 80 seconds.

The most popular shareware Internet Relay Chat program is *mIRC*. To use it, you must first set router port forwarding to 113 for the IP address of the PC you will use it on. You can also change *direct-client-to-client* (DCC) settings to a range from 1024 to 1030.

DirectX and Other Drivers

Modern Internet games are multimedia experiences. The Windows operating system provides a standard software platform that serves as the foundation for many of them, called DirectX. These routines control and expedite video and sound production on your PC. If your game requires DirectX, it will probably load it from the installation CD along with the other game files, but Microsoft periodically updates the package, so it's a good idea to check their web site and get the latest version. You can find it at www.microsoft.com/windows/directx/downloads.

In the same way, your PC's sound and video hardware needs software drivers, and these may be updated from time to time. Some revisions will run best with corresponding versions of DirectX, so you should check the manufacturer's web pages and download the drivers that match. They are free. You may be surprised at the improvement that results.

Nonroutable Protocols

IP is not the only method for transmitting computer data over a network, but it has evolved as the standard way for sending over the global Internet. Some other well-known, old-time favorites are AppleTalk, *Internetwork Packet Exchange* (IPX), *Xerox Network Systems* (XNS), and *NetBios Extended User Interface* (NetBeui). You can continue to use these on your LAN because your Wi-Fi router can bridge them from one local port to another, but the packets lack enough information to be routed to destinations across the Internet. They simply weren't designed for that.

Tweaking: The Need for Speed

If applications such as file downloads seem below par to you, you can check to see if your PC configurations are optimized for Internet use. You may profit by tweaking or making minor adjustments to some of the parameters in your Windows registry. For example, one common problem is that by default your PC will send packets up to a presumed maximum length of 1,500 bytes. But if your ISP uses a PPPoE connection, the maximum packet size is 1,492 bytes or less. Your PC will compensate by unnecessarily fragmenting, or breaking packets apart, to make them fit. When the same payload data is spread across more packets, transfers slow down.

A free test program is available at www.dslreports.com/tweaks, which will download a file to your PC, time the transfer, and then suggest registry adjustments according to your operating system. You may be able to perform these yourself using the Windows program regedit, but care is required. Other commercial programs such as Tweakmaster will do this for you automatically. It is a good idea to back up your registry first in any case.

In this chapter, we learned how to fall into a bit bucket and come out smelling like a rose. The next chapter is more pleasant, as it deals with ways to use a working Wi-Fi to improve your social life, your bank account, and perhaps even your shooting eye.

CHAPTER 7

Wi-Fi for Fun and Profit

Congratulations, my boy. You've taken your first step into a larger world.

—Obi-Wan Kenobi, to Luke Skywalker

Modern computer games have evolved into grand virtual contests between many Internet-connected warriors. The speed of your Wi-Fi network will bring you many other enhanced capabilities, such as mobile video and audio streaming. The same networking tools commonly used by big businesses, such as video conferencing, real-time application sharing and email can make your own home office competitive too.

So Much Music, So Little Time

Until Wi-Fi, Internet listeners had to come to the media, which meant being stuck at a desktop workstation. With Wi-Fi, media can follow the listener around to the kitchen, the living room, or the workshop bench. Manufacturers are introducing products that are even more mobile and designed with the assumption that they will be used with a wireless Internet connection. These "tablet PCs" fill the gap between laptops with keyboards and *personal data assistants* (PDAs) with tiny view screens.

It may be some time before increased bandwidth enables your PC to compete with your TV on an equal basis, but plenty of bandwidth exists today to make your laptop into a superior radio.

World Wide Radio

Thousands of broadcast stations across the globe now stream their audio on to the Internet. What's more, some monaural AM stations simulcast (or *netcast*) a stereo signal. In this way, a listener on a distant continent will likely get better sound from his or her Wi-Fi laptop than a listener in a car one block from the station's radio transmitter. Netcasting enables you to get news and weather from your old hometown or fetch foreign-language broadcasts in real time from Madrid or Seoul. It enables small stations to compete with large ones because part or most of their audience is global. To Internet listeners, no difference exists between a 250-watt, daytime-only broadcaster and a 50,000-watt, clear channel regional broadcaster. You can pick up specialty stations that wouldn't survive in a single metropolitan market but thrive on a base of Internet-aggregated listeners. How about radio-free Talkeetna from Alaska, for instance? Do you like your rap music in French? Stations in Quebec will send it to you for free.

Wi-Fi Moveable Music A Wi-Fi-capable laptop can be carried around the house and yard, requiring no special antennas or yearly fees, as the XM (satellite) broadcasts do. Wi-Fi mobility enables laptops to be companions just as the first transistor radios provided listener convenience back in the sixties. But Wi-Fi radios give listeners far more control over what they hear. Specialized PC jukebox programs play and manage netcast streams from many web sites. They enable users to save source URLs and to access frequently played streams with the same ease they enjoy with a car stereo's buttons. As an extra bonus, after you choose a webcast station, a browser window may pop up, showing accompanying visual information such as

- A picture of the station, the disk jockey, the artist, or album
- Text information about the station or artist
- Possibly, a live webcam view
- Upcoming play lists
- News about the recording industry or popular groups of the same genre
- Interactivity—player buttons that enable you to send a smiley or frowny face back to the originating station for an instant vote on what's being aired
- Email submittal forms
- Real-time request lines

Some streamed audio is sent using proprietary formats, such as RealNetworks' RealAudio, or Microsoft's Windows Media formats. Many others, like the free-download Winamp program, use the almost universal and free MP3 format.

Radio Jukeboxes Many of the sources available to you are radio on demand that forego powerful transmitters and antenna towers. They exist solely to feed specific users via Internet audio streams. Two examples of free-download Internet radio tuners are Winamp and SonicBox, which together offer over 25,000 sources. Unlike analog broadcast stations, these display in text the item being transmitted moment by moment, sometimes the number of listeners (3,000 to 0), and the *bit rate* (quality) of the signal. For some talk-only outlets, a 56K modem link is satisfactory for listening

A Top-rated Internet Radio Station

C. Bryan Miller is the "Internet savant" for WOXY in Oxford, Ohio: "Arbitron has webcast ratings that measure the audiences of online broadcasters. Last month we did about 145,000 *aggregate tuning hours* (ATH or total time spent listening) across all of our web streams. According to their chart, we rank as the most listened to alternative format webcast on the Internet and the twenty-third most listened to webcast overall.

"Our stereo feed comes off the broadcast console and runs through a pair of Gentner Audio Prisms for frequency response leveling, compression, and limiting. The signal is split. One goes out to the FM transmitter and the other comes over to the computer room. Even if there's a failure at the transmitter, the Internet streams are unaffected. The feed then runs through a four-channel stereo distribution amplifier directly into each of the encoder PCs, two of which you can see at the bottom left of the photo. (See Figure 7-1.)

"They all send single streams to servers around the country that handle the distribution and serving. We don't host anything in house because of the massive bandwidth requirements. Our main servers are located in Denver with gigabit Ethernet connections to Level3, AT&T, and Qwest. The RealAudio stream goes to AOL's server farm in Virginia. We peak out during the weekdays at about 30 Mbps outbound throughput.

"I've been thinking about setting up a wireless home network," Bryan added. "So far I've waited because I live in an apartment building and I'm a security freak. I don't think my neighbors are war drivers, but who knows? Plus, my apartment isn't huge and everything is within easy CAT5 reach. But wireless, especially with net radio, is soooo sexy!"

Figure 7-1
Internet engineer C. Bryan Miller in the WOXY computer room. Their music is fed from here to the globe via Shoutcast, Windows Media, RealNetworks, and MP3 streams. Photo by WOXY.com.

because that's probably what's being used to send the talk to the Internet in the first place. A big investment in equipment is not necessary to get started in this endeavor; all it takes is a sound card, specialized software (*shoutcast.com's*, for example), and some kind of link to the Internet. Naturally, a high-speed broadband link is best.

To get near-CD-quality stereo, you'll specify *digital subscriber line* (DSL) or *local area network* (LAN) data rates when you first configure your digital radio. Station listings can be organized by their type or by their popularity. After you assemble a list of preferred hosts on your own, you can organize them however you want. The players, which can be downloaded for free, enable you to search by type or even by a particular song. Some hosts gather an audience by sending several different types of music on several different channels at the same time, with 20 or 30 listeners per channel.

What You Can Hear The Internet radio tuners function like static-free, infinite-range, stereo, short-wave radios with mile-wide tuning dials. The most popular type, or genre, seems to be techno-trance, which sounds like something that the Jetsons would choose if they were into disco and listened in an echo chamber. The radio specialties you would normally expect are there, such as all-news, all-sports, jazz, country, folk, classical, and so on, but the rest is an unpredictable grab bag. If you are the adventuresome type, you can easily spend an entire evening sampling the fare a few moments at a time. The array of audio sources quickly illustrates that the Web is truly worldwide. During one late night session, we heard

- A trio gathered around a piano, obviously inebriated, singing "Mary Had a Little Lamb." Afterward, one of the chorale recited the seven words you can't say on broadcast radio.
- An angry man who asserted that the United Nations and CIA were organizing an invasion of Arizona, so that our wives and children could be forced into concentration camps.
- The Grateful Dead. All day, every day.
- Old-time radio dramas, 24-7.
- Old-time religion, new times Christian rock, and the Book of Mormon.
- Smarmy sex talk.
- The world's largest collection of 78 RPM records from the twenties and thirties.
- Police and fire and aircraft tower radios from major cities and minor towns.
- Pop music in Russian, Israeli, Chinese (Cantonese *and* Mandarin), Spanish, Persian, Japanese, and, incredibly, Esperanto.
- Ambient (background) sounds from Thailand, Vietnam, and Nepal.
- Free and legal downloads from bands who'd like you to buy their CDs.

We heard no commercials, except for an identifying announcement that plays when you first select the source stream. Internet streaming is an ideal medium for college radio stations. Many of them cannot afford FCC licenses, and nowadays almost all dormitory rooms are wired or wirelessly connected to the Internet. It's a small extra step for them to provide their streams to alumni across the globe.

Most laptop speakers are tiny, so you may want to consider buying PC speakers or running an audio patch cord from your PC to your stereo after you find a few stations that you like.

A Vanishing Breed? It is unknown how many of these offerings will be "on the air" in years to come. The SaveInternetRadio.org web site says that "tens of thousands" of webcasters are threatened by the *Digital Millennium Copyright Act* (DMCA), which was passed by Congress in 1998. It granted record companies the right to collect royalties when their copyrighted works were played via digital media, including Internet radio and satellite radio. Traditional radio stations have always paid royalties, but only to composers, not the artists or record labels, and terrestrial broadcasters pay a much lower rate on the theory that traditional airplay serves as a promotion for record/CD sales. But Congress heeded the claims of record labels that Internet streams cut CD sales and subsequently classed streams not as broadcasts, but as *performances*. The distinction meant that webcasters would not be charged a percentage of their income, but instead will pay a per-listener rate. Small webcasters typically don't have advertising support. The mandatory fees, payable retroactively to 1998, will in theory put most of the small-timers out of business. In addition, each webcaster must report

- The name of the service
- The program
- The date and time that the user logged in and out (the user's time zone)
- A unique user identifier and the listener's country

As a Wi-Fi Internet radio listener, you may be unwilling to identify yourself to the music industry. One form of Internet broadcasting that will almost certainly replace the free outfits will be larger, fully commercial enterprises, such as DMX Music, which already provides commercial-free music programming to digital cable and satellite television customers. Their 12-channel net radio service is called the DMX Music Playground. They've announced that they also have entered into an agreement with Sony Music to sell downloads in Windows Media format. Most of those can be burned to CDs.

A Record Player on Steroids

Bandwidth limitations discouraged audiophiles from trading music until 1997 when the MP3 compression standard made it possible to squeeze hours of tunes into minutes of transmission time with relatively little loss of fidelity. MP3 is short for *Moving Picture Experts Group, Audio Layer III*. A typical MP3 compression ratio is about 10 to 1, which yields a file that is about 4MB for a 3-minute track. That's small enough to email or store on your hard drive by the dozens or hundreds. These compressed files let you use your PC as though it were a dedicated MP3 player, like the Diamond *Rio*. A laptop is physically larger, of course, but has a comparatively huge capacity, and new songs can be easily downloaded to it on the fly via the Wi-Fi interface. With a PC so equipped, you can do much more than just play music. You can search for tunes on the Web using specialized search engines.

Copyright Fights Since the Internet is an ideal way to share data, it was only a matter of time before someone developed a web-based method to swap music files globally. But the music industry saw this, and Napster in particular, as a deadly threat to their fees and royalties, and asserted in court that the word *copyright* means what it says. Napster was shut down by a lawsuit in 2001. The site was quickly replaced by software that opened up individuals' hard drives to enable Internet copying directly from one user to another without an intervening or coordinating agency.

The big music pirates moved their operations offshore in 2002 (one of the largest was in China). This prompted the recording industry to try new, aggressive, counterstrategies. The *Recording Industry Association of America* (RIAA) sued major *Internet service providers* (ISPs) in an attempt to force them to shut down access by their customers to illegal sites wherever those sites may be located.

In October 2002, the RIAA filed suit against Verizon, claiming that the ISP should provide the real names of domestic users whom the RIAA believes to be sharing copyrighted material. The RIAA attorneys stated that they want to notify those individuals that such copying is illegal. It is a major test case that pits the privacy rights of individuals against copyrights of content owners. The RIAA also filed suits against web sites that distributed sharing software to individuals, saying that the practice is tantamount to the actual illegal act. The ISPs are not enthusiastic about taking on the unprofitable role of Internet traffic cop, but it is easy to understand why the music industry is so concerned. In 2002, the sales of blank CDs *exceeded* the number of prerecorded ones.

Music on Tap Some performance artists are bypassing traditional marketing channels by offering their music as paid downloads from their own web sites. No other way really exists for hearing some of them, unless they happen to be playing near you. The music industry is struggling to build a new business model for the post-Internet era, but legal versions of Internet music distribution are already filling the gap. Two examples are Verizon's Rhapsody and Earthlink's Digital Music Center. Rhapsody costs under $10 per month and provides access to recordings from 5 major music companies, BMG, EMI Recorded Music, Sony Music Entertainment, Universal Music Group, and Warner Music Group, as well as from over 85 independent labels. That's almost a quarter of a million tracks. Subscribers can listen to a Rhapsody Internet radio station or build and share customized playlists of their favorite songs.

Digital Music Center sells users the right to download up to 50 songs per month for $9.95, or 100 tracks per month for $17.95. Earthlink is actually reselling music from FullAudio, a privately held subscription service. FullAudio buys licenses to distribute music from several major music labels. It offers more than 75,000 tracks. Each user can legally share a downloaded tune among up to three PCs. Needless to say, Wi-Fi makes that a trivial task.

You may be puzzling over how to remember that many songs and find them quickly. The familiar CD player has evolved in the past few years into a digital music management system. These jukebox programs

- Drive your CD burner (or your DVD burner, if you think you need six Gigabytes worth of storage).
- Regulate volume levels.
- Play tunes by artist, album, or any criterion you determine, including randomly.
- Print CD labels.
- Back up your music files.
- Manage the audio streams previously mentioned, as though they were recordings.

Wi-Fi completes the almost effortless link between Internet music servers and mobile fans who have taken to the technology in droves. In the future, look for digital music players that come pre-equipped for wireless Internet access, as PDAs already do. Phillips Electronics NV, for example, is planning to build Wi-Fi into its stereo systems. The market for online music is assured. The all-time winner for search engine requests is the word *sex*. The only word to briefly challenge it for the top spot was the word *MP3*.

PC TV

Although its quality has a way to go before it equals digital television, webcast or streaming video is also growing in popularity because it is video on demand. All the major TV news outlets have web pages that feature streaming audio and video news stories in the most popular formats. On a Wi-Fi laptop, you do not have to wait for news on the hour. It comes when summoned, and you decide for yourself what you want to watch.

Most streaming videos are made for the lowest common throughput, or telephone modem links, so they seem about the same over a high-capacity Wi-Fi/DSL channel. However, they will start to play faster and won't pause or break up as frequently. As more viewers get broadband, the source quality will catch up. Video feeds are labeled by their bit rate. 56K is typical. Occasionally, you will see one at 128K or even 300K. Your home Wi-Fi net has more than enough capacity for higher-quality feeds when you can find them.

The same companies that offer audio management programs, such as RealMedia, offer video jukeboxes as well, usually free for the downloading. Some provide content as well. RealPlayer, for example, is set up to point you to free sources from the moment it is installed. These files come from entertainment companies that want listeners to sample their wares and then presumably buy more of them. They include movie trailers, music videos, and computer game demos.

Look at Me!

The videophone heralded in the fifties has finally arrived, and it is not expensive to buy, nor is it overly complicated to set up and use. Unlike your television set or the video servers described, your PC can also be used to send pictures *out* to the Internet in multiple ways:

- *File Transfer Protocol* (FTP) uploads of webcam shots to your personal web page
- On-demand full-motion files
- Live two-way video conferencing
- Video broadcasting
- Email with pictures attached, often with only a short delay

Video Capture Hardware Fortunately, you won't need to buy a Porta-Cam to start sending pictures, although if you already have one you are a big step ahead toward webcasting. A variety of inexpensive PC cameras are available that plug into your PC's *Universal Serial Bus* (USB), as shown in Figure 7-2. Some come with optional parallel port adapters if you have a spare printer port but no empty USB sockets.

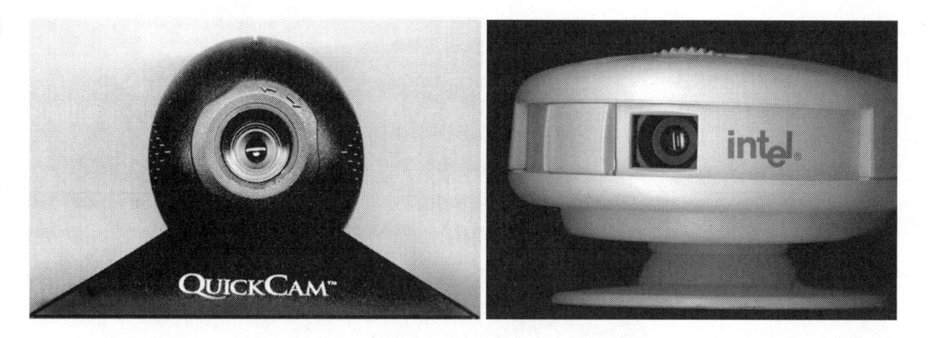

Figure 7-2 Two USB color PC cameras, Logitech and Intel. Photo by Daly Road Graphics.

PC cameras are about the size of your computer's mouse, are designed to work well at indoor light levels, and focus down to a close range. The light sensor inside is known as a *charge-coupled device* (CCD) and they are very resistant to burnout if inadvertently pointed at a bright light source. The level of detail, or *resolution*, is predictably a function of their price, which can range from $25 to $500. The more expensive models have their own web servers and transmitters built in. The median models sell for less than $100, are simple and durable, have built-in microphones, and can be left on continuously. They make a useful peripheral for any PC, but when plugged into a Wi-Fi-equipped laptop, the cameras assume a new dimension of usefulness, because they enable you to send a real-time picture from anywhere.

A PCI-based TV tuner card is another way to connect a VCR, PortaCam, cable or broadcast TV, or another video source to a desktop PC for $100 or less. These cards enable you to watch TV on your PC, capture video pictures to files, make short movies and store them, and also use a higher-quality camera for video conferencing. VCR-quality cameras make a visible difference in the picture you transmit, but they may not be appropriate for full-time use as webcams or Internet security cameras. After weeks of continuous use, their autofocus motors will wear out.

Many newer digital photography cameras come with video takeoff cables, so they can be used as a picture source with these TV tuner cards. Some video conversion peripherals now come in USB-compatible form factors, so they can be used with laptops as well as desktops, and they can be relocated from PC to PC. Small, wireless spycams have been heavily advertised on the Internet, usually on bothersome pop-up web pages, but their images are viewed from a broadcast TV set. You can use these as PC video sources as well, but you must either buy the vendor's video-to-data conversion device at an extra charge or use your own video tuner card for the conversion (see Figure 7-3).

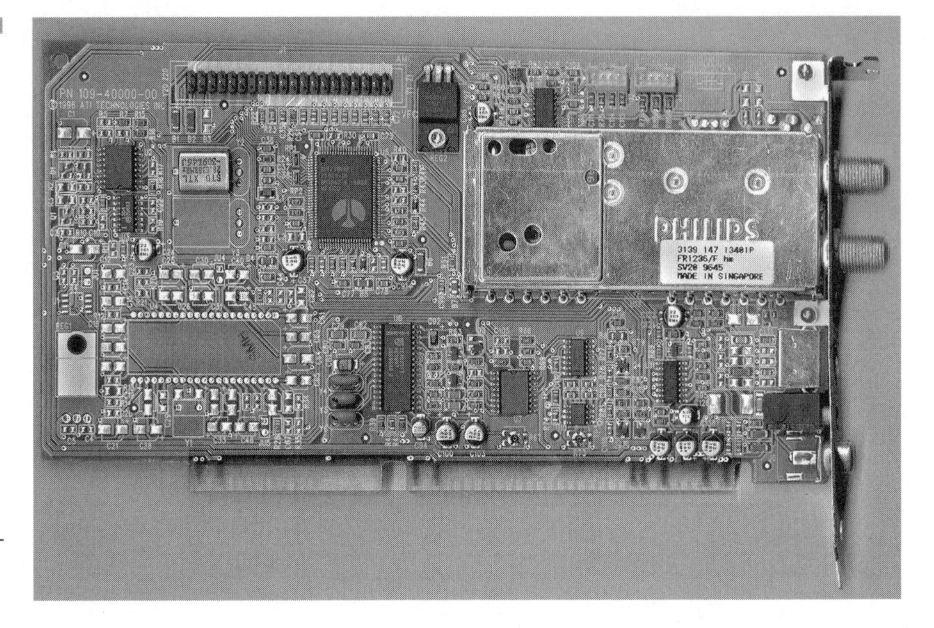

Figure 7-3
A TV tuner on a card. This model (from ATI) only works with other video driver cards by the same manufacturer. It can connect to roof antennas or cable. It also tunes in FM radio stations and can send a picture to a TV set. Closed captioning can be saved to a text file. Photo by Daly Road Graphics.

Webcam Software A webcam is a video camera that periodically takes a snapshot and sends it to a web page where any other cybercitizen can see it. Thousands of them are used all over the world, pointed at everything imaginable. Some monitor goldfish bowls, some students wear them on their heads, and some provide vistas of scenic harbors or not-so-scenic anthills (see Figure 7-4). Now you can get in on this act either for fun or to augment the security of your home. Several inexpensive software packages are available that automate the process. You can download some of them for a temporary trial.

Webcam software makes it easy to publish your pictures on the web by:

- Capturing a picture at predetermined intervals or on a schedule you determine.
- Some will alternate between video sources, if more than one is available.
- They can even embed your picture in a web page; append a caption and the date and time.
- They automatically connect to a web server via a dial-up or DSL line.
- They then automatically upload your picture file or web page using FTP.

Figure 7-4 Wi-Fi webcam software enables you to show your home to the world from practically any vantage point. Photo by Daly Road Graphics.

The superiority of a home broadband link becomes very apparent when you use it to upload webcam photos. Webcams transmit frequently, perhaps once every 10 minutes all day long. A denial of telephone service by a telephone modem at that interval, plus sleepwalking through a 56K upload, would surely tax the user's patience. With a full-time Internet link, your PC can repeat this cycle constantly and quickly in the background, unnoticed by other home net users.

Webcams used to be bolted down, but if yours is attached to a Wi-Fi laptop, you can use it wherever an electrical outlet is available to keep the batteries charged.

Wi-Fi Private Eye As with webcams, Internet-based security cameras have been available for some time, but were not suitable for home use. Industrial versions are dedicated to a single task, mounted in a fixed location, and built to expensive specifications. Home Wi-Fi webcams are cheap and useful in many other ways. Even inexpensive home PC cameras come with bundled security programs. These utilities include a facility for waiting for a picture to change before a frame is automatically captured. In other words, you can put a camera on your living room and nothing will happen until someone walks in front of it. At that time, your laptop can be programmed to take one or a series of pictures and copy them to a server. A PC so equipped can even send an email message to designated addresses and attach the photos to the email. When you combine Wi-Fi's capability to wirelessly transmit images with the small size of a laptop, you have a patient security camera that you can place anywhere, provided that power exists to keep it going. It can be set up to notify you of an intruder while you were on a world tour. You could pick up the snapshots with your email when and wherever convenient.

Movie Files The software provided with your PC cam enables you to use it like a movie camera. You can make motion files and then edit them with simple programs such as Microsoft's Movie Maker or Apple's QuickTime Pro. The latter costs $20, lets you convert between common video formats, and enables crude cut-and-paste edits. You can use it to convert audio files from the Apple to the Wintel world and back again as well, and it has options for choosing the quality of the result. You can upload finished files to your web page so that the world can see your artistic talent. You can even use your PC cam to take a series of single frames at intervals and string them together into a time-lapse film, if you're patient enough.

In any case, the most important thing to remember about movie files is that they are very long. If you pick a fast frame rate to gain smooth motion, a large picture size, and a low compression factor with high-quality stereo sound, you will easily overflow your entire 10MB of personal web space with just one file.

You're on the Air! Times may occur when you want to send a video to many viewers, perhaps to show off a new baby to all the relatives at once or to have a family reunion. Certain web sites will take a video stream from a single source and broadcast it to the world. One of these is WebcamNow (www.webcamnow.com), which provides a free Internet video broadcasting service (software and hosting) for personal use. You can send motion pictures from almost any PC camera or video capture card. All you have to do is apply for an ID, determine which online "community" you will belong to, and then download their software. Two-way communication is possible with your audience through text and voice chats. The audience can see these Webcam broadcasts at a maximum frame size of 320 by 240 pixels. The pictures are compatible with any current web browser that supports *Java*. You can also set up a link to these broadcasts on your own web page by editing in a Java applet, which can also be downloaded for free.

You may wonder why there is no charge for all this. Your picture updates only once every 10 seconds, so it appears as a series of still photographs. If you want something that looks more like a movie (at one frame per second), then you must pay for the greater bandwidth.

Bang! You're Dead!

A killer application is a function so attractive that users will buy hardware or software just to be able to participate. One of these killer apps is networked computer gaming, in which participants are allowed to kill each other—at least, virtually. When your home network is set up and running, you potentially have an arcade to equal or surpass anything at the local shopping mall, because in a Wi-Fi home you can have competitors at several game stations. Preset function keys enable them to interactively hurl insults or commendations at each other as well as laser blasts. Homebrew gaming is better because

- After the initial investment, you don't have to keep putting quarters in.
- The kids won't wander through stores afterward looking for places to spend your money.
- You can keep an eye on them without seeming to.
- If you play too, you'll seem cool. If you actually win, you'll be way cool.
- It is more fun than playing against a PC because machines don't mind losing.

Game playing is actually a real-time simulation with 3-D modeling, which is the most computer-intensive application imaginable. Even so, you don't always need a $2,000 top-of-the-line number cruncher to play. Every game has minimum hardware requirements on the box. You may be able to slide under the bar by decreasing your video resolution, but usually not. Most new PCs have an excess of horsepower anyway. For your "war den," remember to designate the machine with the most horsepower as the host. Actually, some games are smart enough to run tests and pick the most capable PC automatically. All the major dedicated game consoles, Gamecube, PlayStation 2, and Xbox, have online games available as well.

Special Gaming Considerations

If your home Wi-Fi net is already running well, then using it to play games is a simple proposition. But if you plan to import LAN warriors and their equipment, you'll have to make some provisions in advance for the folks who will be toting in their own hardware. That means a wired LAN drop and an empty, always-on AC outlet for each immigrant. If they are using a Wi-Fi-equipped laptop, they will need the power supply (sometimes called the *power brick*) for it and a real joystick. Participants should test their equipment at home before importing it, and as the home SysAdmin, you'll have to loosen your security enough to let them in. They will need unique *Internet Protocol* (IP) addresses so they don't clash with yours.

One final detail: Everyone needs to run the same version of the game(s) and, if necessary, the same databases. The game's version number is usually displayed when it first boots up. Prestaging for game day is as important as for a business presentation. Test first whatever you can. You don't want to hold up the war while one combatant fiddles with wires.

If you don't want to pay for multiple copies, free multiplayer games can be downloaded from the Internet. We'll show you where to find them later in the chapter. These downloads are fairly long, so it may take a few minutes for them to complete. You can speed up the local distribution by saving them first to a file on a local PC, preferably a hardwired one, and downloading to the others from there.

New Worlds to Conquer After you've vanquished everyone in your house and on your block, you can easily find new worlds to conquer in cyberspace. Hundreds of thousands of eager combatants and teammates are out there waiting for you. Many fans of Internet-capable games have formed planetwide conferences or communities, with specialized support web sites that provide advice, patches, databases, and shortcuts (sometimes called *cheats*). The *first-person shoot-'em-up* (FPS) Doom, for example, has a downloadable database that replaces its armor-plated soldiers with shotgun-toting characters from the Simpsons cartoon show.

Making the switch from single player to multiplayer may take some getting used to. Remember that the instant response you expect from your own PC or other terminals on your Wi-Fi network may slow considerably when you are going head to head with a person or crowd on the other side of the globe. If you look at your DSL modem during normal operation, you will see the *enet activity* indicator flicker from time to time. When you are in a game, however, it will be on solid. Huge amounts of data must be relayed from your PC to the others by way of the game server, as it continuously updates a complicated situation. A broadband Internet connection makes multi-player gaming much more enjoyable. As with conferencing, you will need to make some preparations.

Hardware and Security Arrangements The first time you try to engage outside your walls, that is to say, your firewall, you may be frustrated. If your home net security is working the way it should, outsiders will not be able to send unsolicited data to your PCs. You are going to have to make some adjustments to your router to add some deliberate loopholes. Unfortunately, setting up the router to pass game data to several home PCs at once is difficult, as multiplayer games use a variety of ports. Different Wi-Fi routers handle this problem in slightly different ways. Every PC that needs a loophole has to have an entry for every loophole needed, and that can add up to more exceptions than your router can accommodate. For example, the game *Battlecom* uses the following:

- *Transmission Control Protocol* (TCP) ports 2300–2400
- *User Datagram Protocol* (UDP) port 47624

Again, as an example, you would go to your router's port forwarding page and make entries as shown in Figure 7-5.

PORT RANGE FORWARDING

Port forwarding can be used to set up public services on your network. When users from the Internet make certain requests on your router, they will be redirected to the specified IP.

Customized Applications	Ext.Port			Protocol TCP	Protocol UDP	IP Address	Enable
	2300	To	2400	☑	☐	192.168.1.200	☑
	47624	To	47624	☐	☑	192.168.1.200	☑
	0	To	0	☐	☐	192.168.1.0	☐
	0	To	0	☐	☐	192.168.1.0	☐
	0	To	0	☐	☐	192.168.1.0	☐
	0	To	0	☐	☐	192.168.1.0	☐
	0	To	0	☐	☐	192.168.1.0	☐
	0	To	0	☐	☐	192.168.1.0	☐
	0	To	0	☐	☐	192.168.1.0	☐
	0	To	0	☐	☐	192.168.1.0	☐

UPnP Forwarding | Port Triggering

Apply | Cancel

Figure 7-5 Opening portholes in a firewall. Rules were made to be broken.

For this example, we have chosen .200 as the local IP address for the gaming computer. Remember that the router hands out local IP addresses when your PCs boot up and ask for them. So, unless you turn *Dynamic Host Configuration Protocol* (DHCP) off and manually assign a permanent local address, your gaming PCs may come up with game ports disabled again. If you are the only one to game with the outside, you can to pick one PC as your favorite gaming machine and tell the router to pass *everything* by using its built-in "passthrough," Exposed Computer, or *demilitarized zone* (DMZ) feature. Then the port range used by your favorite game won't matter. Of course, when you do this, your game PC no longer has the router's firewall protection either. (This caution bears repeating.)

Latency *Latency* is the network-induced delay between an action and a response. It's unnoticeable when browsing, but in a competition that hinges on quick reflexes, it can be very aggravating. Some have compared it to driving a car with a rubber steering wheel.

One way to put a number on the nuisance is to use your PC's ping program. It sends out a packet, starts a stopwatch, and measures the time required for an echo to return. For a telephone modem, this round trip can take 150 to 250 milliseconds. On a cable DSL connection, it can be as brief as 20. Times of 30 to 70 are typical on a Telco DSL circuit. If it is higher than 350, you will feel as though you are trying to steer the Queen Mary with short oars.

If your ping times are consistently high, but overall speed seems okay, you might want to question your ISP about the route data takes from your home to the Internet. Even though some ISPs are local, they actually resell a service from a national provider who actually owns the Internet gateways. To reach them, your data may have to bounce across the continent before it gets to a nearby game. Luckily, some latency problems go away if a temporary router failure or poor network planning caused them.

You can sometimes diagnose the problem on your own by using a variant of the ping program previously mentioned. Many come with a *traceroute* feature, with gives you a list of all the intermediate hops between your machine and a target server (refer to Chapter 6, "Can You Hear Me Now?"). Pinging each in turn will give you a clue as to which is the bottleneck. If too many exist to ping this way, then you have found your problem, because each intermediate router contributes to the latency total. There may not be much you can do about that except for voting with your feet, or in other words, fire your ISP and buy a connection from someone else.

Latency is so important that it is often displayed on a multiplayer game's scoreboard. A good player with a slow link will usually beat a novice with a fast link, but all other things being equal, a slow link will get you shot early on. If you are playing with someone else on your home net, even over a Wi-Fi link, latency will be practically zero. Latency can change while a game is in progress, as players join or leave, or congestion on your Internet link changes. Also, when your opponent has a slow link, your delayed responses from them will slow down the game. Because all home Wi-Fi players share the same access point bandwidth of 11 Mbps, you may want to consider putting as many of them as possible on the wired access point ports, which do not share bandwidth and go 10 times as fast.

Gaming is one of the few applications where users might approach a Wi-Fi connection's saturation point. A fast link is most important for a locally based game server, because all the others will depend on it. Even at home, some players will have lower ping times than others, depending on their hardware. Some games provide advice on fine-tuning to reduce latency. Also, you may want to shut down other background programs that eat up bandwidth, such as chats.

Game Lag Another common frustration in Internet play is known as *lag*. Lag is due to momentary network congestion, the same phenomenon that causes video streams to freeze and music to pause. It can be very enervating in a video game, because your character momentarily has his or her feet glued to the virtual floor while the world zips by. If someone else is lagging, they may seem to be petrified one second and then zoom off like the Flash. That's the result of the game server catching up with late-breaking data.

Helpful Gaming Sites Many games enable you to send chat messages to other players. Refrain from doing so unless you have something to say, because they came to shoot, not shoot the breeze. Game forums are a better venue for conversation. Here are some popular multiplayer game resources:

- www.gamespy.com
- www2.station.sony.com/en
- www.happypuppy.com
- www.gamespot.com
- www.ezgamez.gofrag.com
- www.adrenalinevault.com
- www.gamesdomain.com

These web sites feature forums, free downloads, *Frequently Asked Question* (FAQ) pages, news, cheats, patches, demos, desktop wallpaper, and links to still more resources.

Have Fun (That's an Order!) Remember the old phrase, "It's only a game." Don't get frustrated. It takes time to learn how to blow stuff up without getting vaporized. If it stops being fun, walk away and do something else for a while. If you find yourself frustrated at 2 A.M., walk away and sleep. It will improve your reaction time and your social life. Also, these games have different levels of skill for a reason. If you drop into a tournament of lightning-fast thumb-jockeys, you will likely get zapped before you can leave square one. It is more fun if you work your way up. You may be able to join in spectator mode, which allows you to see the world through another player's eyes. That's a good way to learn the ropes. You may get mercy if you log in with the name "Player," which is a factory default for *newbies,* or newcomers. But don't count on getting a break. Eventually, you will want your very own *nom de guerre*, or *nick.* Try to be original and brief, and stick with whatever you pick. For someday, you too may be a legend.

Wi-Fi Telephones

One application that seems a natural for Wi-Fi is Internet telephony. All long-distance telephone calls these days are digital, in that the audio is digitized before it is put on the telephone company's networks. It is possible—even easy—to send audio over the Internet or even to videoconference with sound, and cordless telephones are a common commodity. It is natural to wonder, then, why we have not seen a dedicated Wi-Fi telephone handset that sends your digitized voice over the Internet, bypassing Ma Bell's long-distance charges. Some of the reasons for the slow start are the same as for the original attempts at wireless networks. Early implementations were unreliable and required proprietary, nonstandard LAN telephones.

Even so, some large companies are profitably using *voice over IP* (VoIP) products available today to carry their long-distance communications over the unused capacity on their corporate intranets. They have opportunities that home businesses do not. National corporations have already paid for a full-time channel that they own exclusively, and they have plenty of traffic between hundreds or thousands of phones on either end, which justifies the initial investment in specialized interface equipment.

In 2001 some firms emerged that specialize in connecting small offices using Voice Over Internet Protocol (VoIP). Verizon announced a joint venture with gobeam.com called Voice Over Broadband, which will provide local and long distance phone service, voice mail, faxes and teleconferencing over a single, combined broadband link. The foremost problem with home Internet Phones, however, is the type of IP data packet they use. They do not have top priority. If some get lost en route, the receiving end will not ask for a retransmit, and an Internet channel usually carries many kinds of data. The TCP/IP protocol consists of detailed descriptions of some 50 different types, each with its own order of precedence through the pipe. For email, brief pauses in the data stream are not a problem, but users notice instantly if a human voice gets chopped up or delayed because a higher-priority packet pushed it aside. The telephone company networks don't have the problem because they are specifically designed to carry voice.

These issues fall into the category of *quality of service* (QoS). Wi-Fi didn't become widely popular until all manufacturers adopted a common technical specification for it, but no such standard guarantees that voice data will go first on your home network. As Gemma Paulo of InStat/MDR research told us, manufacturers are waiting for one.

"The QoS for Wi-Fi 802.11x has not been standardized yet," she said. "It is crucial that voice be prioritized over data when both are being passed on a wireless LAN. Task Group E of the IEEE 802.11 Working Group is currently working on this standard.

"The Wi-Fi handsets available from Spectralink and Symbol retail for around $700 each and generally go into three main verticals: retail, healthcare providers, and education. These handsets are currently too expensive for the general consumer. Once the QoS is standardized, then more vendors are expected to get into this market. This may drive prices down," she predicted.

Even so, In-Stat/MDR predicts that VoIP handset sales are expected to surpass half a million units by 2006. Until then, you are left with the option of using a softphone solution from your PC.

Portable Softphones

As with video conferencing, using a Wi-Fi laptop is considerably more convenient because you can make your calls with the same ease as with a cordless phone. Some examples of software telephones are gotocall.com and a dozen others are listed at www.tucows.com/phone95.html. Some require that you prepay to set up an account. Afterward, you simply open a browser window, which lets you enter the target number and it displays the duration and cost of each call, which is typically 3.5 cents per minute in the United States and 4 cents to the United Kingdom. Their facilities allow you to place calls to regular, non-IP telephones. If the party you wish to talk to has a PC, you can use teleconferencing software for voice, as we describe later in this chapter. But sound quality is generally superior with softphones.

Configuring Your Router for an IP Phone To use these programs, you will have to make adjustments similar to those previously mentioned for video conferencing and Internet gaming (refer to the previous examples) or your firewall will shut off the connection. As an example, the Net2Phone program uses three IP ports. You would need to go to your router's port forwarding page and open UDP port 6801 for your telephone PC. For the other UDP and TCP ports, open anything in the range from 1 to 30,000. You can avoid this if you want to invest in a router/firewall specifically designed to enable VoIP (see Figure 7-6). Of course, if all else fails, you can open up your phone PC or laptop by putting its IP address in the router's DMZ category, meaning that any outside packets will get through to it. IP phone providers also provide technical support to assist you with these details.

You can buy several accessories to make your calling easier. A good headset helps and you can also buy USB-compatible speakerphones and handsets that plug into your Wi-Fi laptop.

Figure 7-6
This router enables Internet phone calls using only a regular, plug-in telephone. Courtesy of Linksys.

Wi-Fi as a Tool for the Job

Throughout these chapters we've described the amusing or enlightening things you can do with a wireless connection to the Internet. If you operate a business from your home, or even if you do corporate work from home, some of these capabilities may have already seemed prospectively useful to you as a means to make or save money. Let's take a look at them in that light.

As Anthony Townsend of NYCwireless points out in his interview in Chapter 8, "Where No One has Roamed Before," working at home was normal for most of human history until big industrial tools like steam engines rendered that impossible. Chances are you won't have to go back far on your family tree to find a relative who lived very close to their work (our grandfather had a metal shop in his garage, and our grandmother sold insurance from her living room). Thanks to portable PCs, the rollout of the Internet to the home, and the modern shift from a manufacturing to a service economy, the term *cottage industry* could easily be updated these days to read *apartment industry*. In 2001, the Bureau of Labor Statistics reported that 19.8 million persons typically did at least some work at home as part of their primary job. These workers reported working at home at least once per week. They accounted for 15 percent of total employment.

Although most corporations have reservations about employees working exclusively from their homes (see Chapter 8's Russell and Chapin interviews), most employees now consider some work from home as part of their job description. The Pew Internet Project's Broadband Report says that a third of home broadband users telecommute occasionally. Over half of them (58 percent) say they spend more time working there thanks to powerful residential Internet access.

It's true that personal interaction between staff is impossible at home and that teleconferencing is not exactly the same thing. But not all of an employee's day is spent talking face to face. A home broadband Internet connection is no different from the one at work, especially if the employee can use virtual private networking (see Chapter 4, Virtual Private Networks) to access company mail and other records quickly and securely. Some corporations will accept full-time work from home if it is a cheaper alternative to rented office space (see Chapter 8's Chapin interview).

More frequently of late, the corporation, the CEO, and the employee are the same person, and the corporate headquarters also doubles as the employee residence. Possibly, like many baby boomers, you have decided to strike out from your corporate cubicle willingly to be your own boss. Perhaps you are putting your corporate skills to profitable use in a home-based business due to widespread downsizing in an uncertain economy. That same uncertainty might be prompting you to start a sideline business of your own just in case. John Knowlton in the online magazine *GoHome.com* says that 2 million new entrepreneurs take that plunge every year. If you are one of them, you will be happy to learn that Wi-Fi continues the dissemination of low-cost, powerful technology from the data center outward and downward to your kitchen tabletop.

Why Invest in Wi-Fi? Manufacturers sell Wi-Fi to big corporations using the argument that it will gain their workers two minutes a day in extra productivity, which by their reckoning will quickly recover the cost of the investment. The same argument logically applies to the *Small or Home Office* (SOHO) as well. Tax laws provide the same incentives and benefits for small businesses as for large ones. When making a living at home, former costs of living paid out from your disposable income assume the status of reasonable business expenses, including

- A laptop computer
- A Wi-Fi PDA
- DSL and ISP service
- Other computer equipment as well as installation and consulting fees
- Hubs, Wi-Fi printer drivers, and the printers
- Wi-Fi access points and *network interface cards* (NICs)
- Business-related software
- Other equipment such as scanners, PC cameras, and headsets
- A second phone line

High-speed Internet access has made it more possible than ever for small-time operators to meet their big competitors in cyberspace as equals and to do so with unprecedented economy. It might not make sense to buy a Wi-Fi LAN to keep your kids happy, but buying one to keep your customers happy is good business sense. One of the arguments for making your home office into a hotspot is to make it like all the other businesses. Within a few years, any business that cannot provide wireless connectivity for a visiting customer (as coffee shops, hotels and airport lounges do already) will be considered second-rate, like businesses that can't deal with credit cards.

Wi-Fi Business Terminals The laptop has become the all-purpose tool of choice for the mobile businessperson. In late 2002, Dell Computers joined Toshiba, Hewlett-Packard, and others by announcing that they would build Wi-Fi into their product line. In that same year, Gartner Dataquest forecast that professional mobile PCs with built-in wireless interfaces would grow to almost 50 percent of total laptops by the end of 2003, and surpass 90 percent by 2007.

The PDA is rapidly ascending to the laptop's must-have status, especially Wi-Fi-capable varieties. Surveys by In-Stat/MDR show that about 10 percent of the U.S. workforce currently uses some sort of PDA or handheld for business purposes. The ability to access the Internet from a PDA is being increasingly integrated into them. About 19 percent of PDAs will offer this functionality in 2002, and this percentage is expected to grow to 70 percent by 2006 according to In-Stat/MDR. Thanks to the interoperability of Wi-Fi terminals, your PDA will work on your home Wi-Fi network as well as at the airport or coffee shop. It is not unusual for domestic entrepreneurs to entertain customers in their homes to meet or exchange information. The ability to wirelessly copy files there will save time for themselves and their customers.

Teleconferencing: The Answer to a Road Warrior's Prayer

A PC's most useful asset is its capability not just to organize information and present it, but to communicate it anywhere quickly, at low cost. You may remember that wireless computer networks were invented to save the time, expense, and bother of business travel (refer to Chapter 1, "Is This Trip Necessary?").

One of the most expensive if not boring activities a small business operator does is drive to customer sites to get or drop off information. In addition to gasoline costs, one faces the added burden of what economists call *opportunity cost*, which is the unprofitable, unrecoverable loss of time. If the Internet can connect vendor and customer, then the time expense of knowledge transfer drops to almost zero. Even at a customer site, it's possible to assemble a network in moments if both parties have Wi-Fi equipment set to ad hoc mode. This enables the transfer of files too large for a floppy, a zip drive, or even a CD-ROM.

If you work for a corporation that requires travel or you work for yourself in an occupation that requires it, you know that the skies are not as friendly as they used to be. The attacks of 9-11 resulted in long lines, intensive security checks, and a heightened sense of anxiety at airports. The economic slowdown that followed made the expenses of business travel less bearable than ever. Virtual conferencing provides a means of meeting a client face to face without spending hours in taxis, cramped planes, rented cars, hotel rooms, and then returning home to days of jet lag. It is not a complete substitute for personal presence, but it is an alternative, and many corporate meetings don't require a real handshake.

Professional services exist that can set up teleconferences for you, using high-quality, full-motion video cameras, stereo sound, plush studios, hot coffee, and facilitators at both ends. But if you use a Windows-based PC you already have one of many virtual conference programs that make it possible to send a live image at considerably less expense from your home office. *Netmeeting* is the virtual conferencing application that comes bundled with Windows, and it is downloadable for free from the Microsoft web page (www.microsoft.com/windows/Netmeeting). The driving software is standard enough to enable it to interoperate with other manufacturers' programs.

A similar program, CuSeeMe (www.cuseemeworld.com), enables up to 12 users to put up windows in a user-defined virtual conference room. It sells for $40. Neither program requires that a user have a camera to participate in a conference or watch the other participants. In addition, videoconference software establishes a multimedia channel to share data of many types with several individuals in remote locations.

Video conferencing is a killer application that is changing the way business is conducted, because the old adage about "a picture being worth a thousand words" is incorrect. Sometimes no number of words can instantly convey as much information as a color video (see Figure 7-7). Wi-Fi networking adds a particularly useful dimension to it because in the past cameras had to remain attached to bulky desktops. Video show-and-tell is much more powerful when the camera, like the human eye, can wander freely and transmit in real time. With a Wi-Fi camera and a little imagination, you can

- Sell real estate to prospects relocating from another city, showing it off inside and outside.
- Meet with scattered salespersons or field engineers from your desk, living room, or hotel room. (Hotels are quickly being equipped as Wi-Fi hotspots.)
- Set up a "video classroom" from a site using real props and live interviews.

Figure 7-7 Show the mechanic what you are talking about with your Wi-Fi eye. Photo by Daly Road Graphics.

- Show a leaky pipe to a plumber and get an estimate.
- Demonstrate how to assemble or repair equipment, and answer questions on the fly.
- Give live coverage from the convention floor (televise events as they happen).

Be Considerate of Others You should be aware that some people are made quite nervous or resentful by the presence of video cameras. Some PC cams are built with trapdoors covering their lenses as much to show that they are disabled as to protect the lens. You might also hang a hat over yours, point it out the window or unplug it, and put it in a drawer when it's not in use. Think very carefully about using a PC camera to secretly spy on someone. The fact that cheap equipment to do so is widely available does not make it less offensive. If it is discovered, your victims will not accept your actions as a joke. Some workplace regulations state that everyone within visual range of a surveillance camera must be warned in advance of its presence. (See "The Last Word" in Chapter 8.)

Conference Sound

Instant feedback in the form of sound makes co-coordinating a meeting much easier. It also allows conferees to make de facto Internet phone calls via conferencing or other software. All recently manufactured laptops have built-in speakers and pinhole microphones, though the latter may not be obvious. If you are conferencing from a desktop PC, however, you must install a sound card for this capability if you don't already have an audio card. A $10 model seems to work adequately for this purpose.

Inexpensive headsets for laptops or tabletop PCs can also be purchased for about $10 and are worth the investment (see Figure 7-8). They have surprising quality and enable you to use your hands to carry the laptop, move the camera, and type on the keyboard. Headsets will also let you turn off your speakers, which can be a severe distraction when you are trying to talk, thanks to a phenomenon known as *cockpit effect*. Headsets also muffle loud background noises such as cooling fans, so you can hear your partners more clearly. Similarly, a microphone close to your mouth (a finger's breadth from the lips is best) makes you more easily heard than a laptop's pinhole mike, which also picks up keyboard clicking and other background noises. This can be a problem for some conference software, which automatically detects speech, and turns the microphone on for it automatically. It may also turn on the microphone for cooling fans or loud office equipment, just as a speakerphone would. For parents who perceive Wi-Fi Internet radio stations as noise, headsets can keep the peace between themselves and their offspring. Headsets also work on boom boxes, portable TV sets, and so on. PC microphones will probably *not* work on tape recorders or VCRs unless they are labeled as using *condenser*-type microphones, which draw operating power from the device they are plugged into.

One at a Time, Please Sound in a videoconference doesn't work exactly like sound on a telephone. As Internet game players can tell you, latency is the built-in delay between the time you dispatch data and the time it arrives at its destination. When combined with the delay of a response, it can add up to a second. This pause can be very confusing. Internet telephone software is designed to minimize it, but not conferencing applications. The simplest way to deal with it is to talk as though you were using a speakerphone; both conferees cannot speak at the same time, nor can one user interrupt another. Speakers must remember to communicate in complete sentences or, better yet, paragraphs, the way ham radio operators do. A simple way to signal the group that you're finished is to ask "Okay?"

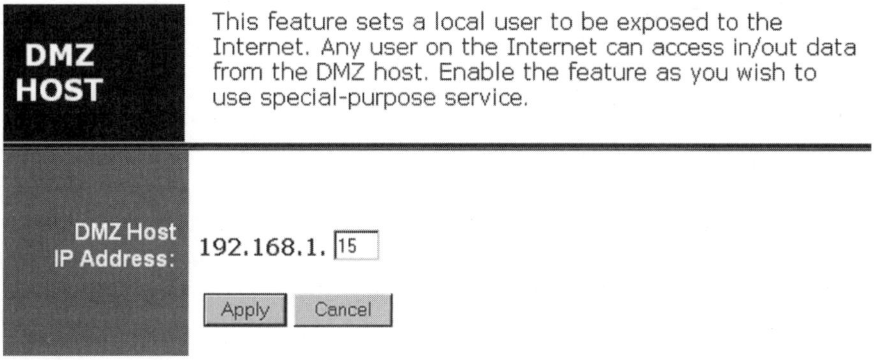

Figure 7-8 Teleconferencing will almost certainly require that you put your conferencing PC outside of your router firewall.

DMZ HOST

This feature sets a local user to be exposed to the Internet. Any user on the Internet can access in/out data from the DMZ host. Enable the feature as you wish to use special-purpose service.

DMZ Host IP Address: 192.168.1. 15

Apply Cancel

Other Conference Features

Seeing your coworkers and talking to them from a distance are spectacular benefits of teleconference programs, but such capabilities are not always the most useful or productive. These teleconference programs swap data between PCs in a number of ways and forms. Some of them can be more productive than a personal visit. Here are some other conference features:

- *Application sharing* When one employee walks to another's cube to haggle over the wording of a document, for example, they will wind up passing the keyboard and mouse between each other. By sharing the document with conference software, they can use their own keyboards and edit the same document while both see the changes at once. The same process works for spreadsheets, CAD programs, or database inquiries. The collaborators can be separated or be elbow to elbow. With this technology, everyone can be side by side in cyberspace.
- *The whiteboard* The whiteboard is a shared pop-up screen that can be used like the traditional classroom blackboard. Conferees can draw multicolored lines, doodles, or diagrams on it, using their mice. In practice, it behaves like a *Paint* program, except that everyone in the conference can see what is being drawn, and they can draw upon it themselves.
- *The desktop* The entire desktop view of a conferee can be shared. That is, partners can see exactly what the sharing user is looking at, including icons and the cursor. What's more, the sharing partner can allow another to remotely control their PC. This makes it possible to remotely troubleshoot some PC problems. Remote maintenance and

patch installations can also be carried out using desktop sharing by a distant technician who types commands, runs tests and views results as though they were sitting at their own machine. The feature can be used for training, as one person shows another how to run an application as though the pair were sitting next to each other. Once desktop sharing is enabled, security becomes doubly important. These dual-control programs allow various levels of login and data encrytion, so that foreigners can take over your computer by invitation only.

This feature is so useful and comprehensive that programs are marketed that specialize in PC remote control. Buyers use them to operate their home PCs from satellite workstations. A laptop user at a customer site can make their home PC initiate a file transfer in this way.

- *Chat* Conferees without sound cards can "chat" to one another. This uses little bandwidth and enables users to make comments while other higher-volume transfers are underway. The text can be saved to a file for future reference. Cut and paste is permitted. You can use it to display exact spellings or quotes to multiple conferees.
- *File copies* Files of any type can be copied from one PC to another or to all other conferees over the Internet, on the spot, using the same drag and drop interface you'd use for a local copy. Most corporate email programs limit the size of attachments to 1 or 2MB, and email may have an unpredictable transfer delay. Those limitations don't apply to real-time transfers, and confirmation of a successful receipt is instantaneous. Before the meeting begins, last-minute reference materials can be distributed over this channel. When a document has been modified and approved by a cybercommittee, the final version can be sent to all the participants before the meeting ends.

Configuring Your Router for Virtual Conferencing

Video conferencing presents some unique problems when carried out over a home router. The programs push your network's data-carrying capacity to its limit. They use a variety of ports, some of which are automatically negotiated when the meeting commences. Some home routers are designed to accommodate teleconferencing requirements, and some will not. You should take that into account when evaluating a potential purchase. If your router does not, you will have to make some changes before you make the meeting.

Connecting to Your Partner(s) The first technical hurdle to be overcome for a conference is the connection, especially through a router, wired or wireless. The conference software producers originally envisioned their users finding each other on the Internet or an intranet at common points known as Internet Locator Servers (ILS). Parties log on, their names are displayed along with other identifying information, and all can connect to the others by personal or conference names. These conference servers act like Internet telephone directories by converting user names into IP addresses. Microsoft stopped supplying this service to the public, but others have filled the gap. One problem in using it is that lots of surfers may see your name and try to connect to your conference, particularly if you are logged in with a female name. (As you might guess, video dating or something more intense is one popular application for this technology.) You might want to configure the program to *not* log on to an ILS server automatically at startup (see Figure 7-9). Some teleconference software vendors provide customers with virtual conference rooms that restrict entry to conferees only.

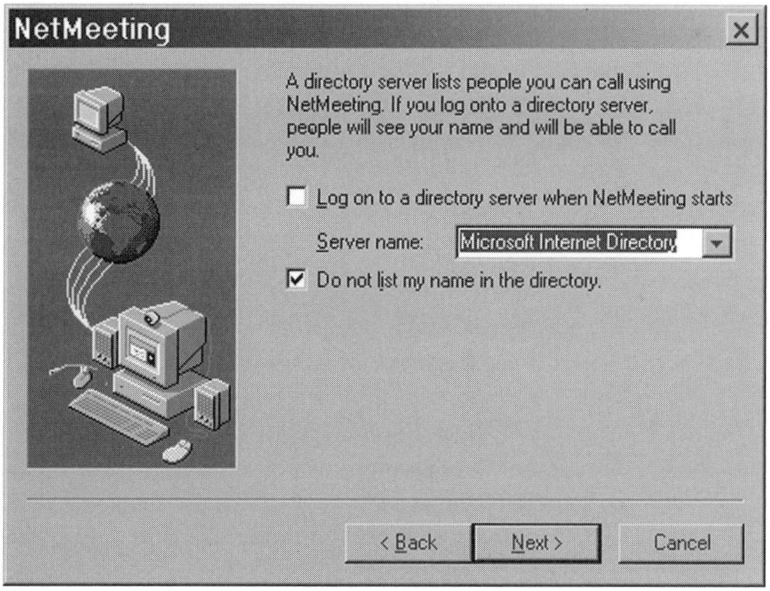

NetMeeting

A directory server lists people you can call using NetMeeting. If you log onto a directory server, people will see your name and will be able to call you.

☐ Log on to a directory server when NetMeeting starts

Server name: Microsoft Internet Directory

☑ Do not list my name in the directory.

< Back | Next > | Cancel

Figure 7-9 Turning off NetMeeting's logon server. You don't need help to reach out and touch someone.

You can bypass logon servers and directly connect to a partner by using his or her IP address if you know it, instead of a text name. Only the one initiating the connection must know the address of the other. If you originate the call there will be no problem. However, your home Wi-Fi router's NAT facilitates the sharing an IP address amongst several computers for *outgoing* purposes only. It is supposed to restrict inbound access from the outside. Your Wi-Fi firewall will therefore hide you from incoming conference calls, as will your PC's personal firewall software, unless you have specifically set them up to pass the caller.

Like the games previously mentioned, some conferencing programs use a variety of IP ports. Only some of NetMeeting's functions can be made to work by using permanently forwarded ports, because the ones controlling video and audio transfers are negotiated on the fly when the connection is first set up. You don't know which ones the conferees will automatically pick, so you can't prepare a door for them.

If you want to do a videoconference between two PCs in your home, this problem does not apply, because both are on the same side (the inside) of the firewall. When you are calling the other PC in that case, you would give the local IP address of the target, such as 192.168.1.15. No special configurations are necessary. In this way, you can use a Wi-Fi laptop as a baby monitor or remote camera.

A measure of last resort can be taken for outsiders, and that is to pass data straight through to your conferencing PC, at least temporarily. As with game hosts, you must put the local IP address of the PC you want to use for NetMeeting in the DMZ or passthrough port of your router (see Figure 7-8). When you do this, the PC is open to any unsolicited traffic from the Internet, including packets from hackers. Fortunately, passthrough can often be turned off again with a single mouse click.

After the host PC's local address has been unrestricted in this fashion, the other party calls by entering the IP address that the router presents to the world, the *wide area network* (WAN) IP. Don't attempt to test the connection by calling the host from a local PC using this address; it doesn't work. If you want to test in advance of a conference, you can do so by using your dial-up connection or another PC that is outside your firewall. You might also check the web page of your router's manufacturer to determine if a newer firmware release offers support for NetMeeting or other video software.

If you are not placing the call, and your router does not have a static IP address, you will have to see what your address of the moment happens to be and communicate it to the caller. You can find your IP address by interrogating your Wi-Fi access point's status page or by consulting the web

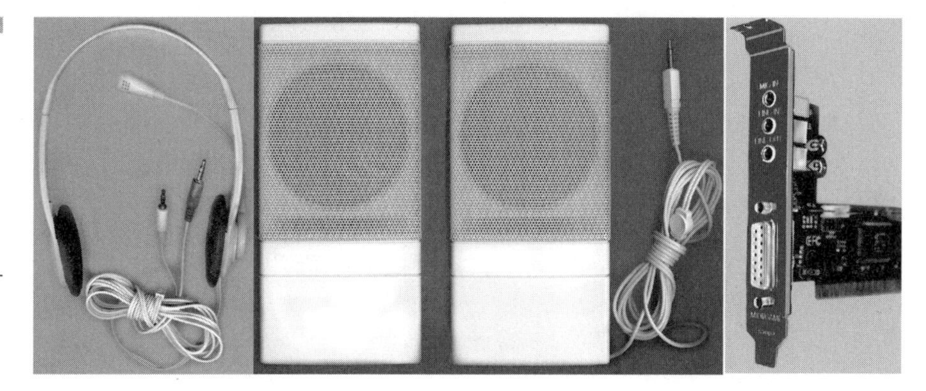

Figure 7-10
A headset, speakers, and a sound card. These accessories for video conferencing sell for less than $10 each. (Photo Daly Road Graphics)

page, www.whatismyip.com. If you use IPCONFIG or one of the other Windows utilities, the address returned to you is the nonroutable local address. After you learn the WAN address, the one your Wi-Fi access point is using to communicate, the simplest method for sending it to your partner is via email or buddy chat. Fortunately, temporary addresses don't change from minute to minute. You may hold onto yours for weeks at a time.

Conserve Bandwidth Before the meeting starts, take a moment to check your Wi-Fi interface's configuration guide to guarantee that your link is running at 11 Mbps. A slower speed may not be noticeable until you put a heavy data load on it. Usually, you can fix a slow link by moving the antenna a few inches. Also, begin the session with your video turned off, particularly if you are going to begin talking at once. Video occupies more of the PC's available bandwidth than audio. Sending a high-resolution picture involves a trade-off between the frame refresh rate and picture quality. If you send a large image, it may interfere with your audio. You'll have to experiment to determine what your limits and needs are. You won't need much detail for a "talking head," but you might for a televised football play. Remember that if your partner is on a slow link, that will seriously limit how much you can send.

Choosing "send a medium-size picture" and "prefer to receive higher quality" seems to be a good compromise. Sending lots of video data can also stall a big point-to-point file transfer, so pause or turn off the video steam when you're not actually using it. ADSL lines are typically much faster downstream than upstream and can appear to choke when simultaneous up- and downloads are required, as they are in a two-way videoconference. It will be even more noticeable to other users of your Wi-Fi network, who may be suffering from the problem or unconsciously adding to it with their own data demands.

Email

One of the simplest tools of a contemporary business is also the most effective. Email is simple to set up, almost free to use, and a great equalizer for small offices. In the same way that the Internet appears to your home Wi-Fi network exactly as it does to a corporate PC, your email to large corporations appears exactly like those from other large corporations. No one can tell from an email if Acme Gizmos, Inc. has 2 or 200 employees. This is as it should be, because most of the time the size of your customer base or gross earnings is less important than your business' reputation, which you build for yourself. In fact, being small may be a competitive advantage because it allows you to react quickly and respond with more flexibility to customers' changing requirements. To do so, you must keep in touch with them.

A Wi-Fi laptop permits you to carry on this important part of your work anywhere in the house and, more recently, in the airport, hotel, or coffee shop. Hotspots outside the home or corporate campus are *day extenders.* They enable you to carry your office with you and carry on your business as you move.

Virtual Private Networking

As the Internet became more pervasive, businesses turned to this new channel as a means to extend their intranets to remote offices. As the Internet rolled out through the last mile into employees homes, companies set up VPNs, or virtual private networks so that workers could telecommute. The number of people who work from their homes has steadily increased in recent years. One explanation is that corporations are closing down rented remote facilities to save money. Of course, many employees willingly work from home as well. In either case, they must send information to and from their corporate offices, and do so securely. Unlike a corporate intranet, however, remote users data must traverse insecure links, including a WiFi home network if one is used.

A VPN is a proprietary network that establishes virtual connections using a public network such as the Internet as one of its constituent links. In the past, corporations had to lease dedicated data lines to distant offices. One advantage of doing so was that hackers were less likely to read the data going back and forth on them. VPN solves the problem of insecure broadband links by encrypting data before inserting it inside an otherwise normal IP packet. After the enciphered payload arrives at the intranet destination, it is stripped out, decrypted, and then retransmitted as though it had originated on the headquarters campus. Outgoing data gets the same treatment when it arrives at a company PC located in a home or an apartment. You will not need VPN for non-commercial endeavors. But if you do need it, in order for your home computer to use VPN:

- It must have a static IP address.
- It must have unrestricted access to the Internet. This means that your WiFi router must permit VPN. Some do not. Choose yours accordingly.
- In the same way, if the Internet connection is established through a proxy or a special gateway, that device must accommodate a VPN.
- You will need VPN software from the corporate Internal Telecommunications office. VPN can be implemented in different ways, and yours must be consistent with theirs.
- You need a login name and password that gets you through the corporate gateway to the corporate VPN server.

Often, a corporation that requires a VPN for many users will outsource the job to an *enterprise service provider* (ESP), which provides corporate employees with software and training. The ESP also sets up a network access or VPN server. These servers do more than translate data. When VPN requests for sessions arrive at the corporate Internet interface, they are proxied to the *Accounting, Authorization, and Authentication* (AAA) server, which verifies the user's function, permissions, and identity.

Generally speaking, here's how a VPN works. After you've set up the VPN software on your home PC, it acts as a virtual network adapter. When you send data to the corporate office, it first makes a stop at this internal device. There, your data is scrambled beyond recognition. IP Security (*Ipsec*) is the most widely used VPN mechanism for securing packets. There are a number of algorithms that it can use to protect the payload. IPsec is made up of three main elements:

- *Encapsulation Security Payload (ESP)* Part of the packet's description, or header and all data is encrypted using the Data Encryption Standard (DES) developed by IBM.
- *Athentication Header (AH)* This mixes authentication data into the payload, guaranteeing that it won't be readable by an unauthorized viewer or altered en route.

- *Internet Key Exchange (IKE)* is used to exchange and verify cipher keys. It is a set of management protocols used to negotiate which encryption methods will be used by ESP and AH, above.

After all this, user data is *encapsulated* in a standard IP packet (the envelope) before it is dispatched to the Internet in the standard manner.

It's possible to put almost any kind of network packet into the IP envelope, including ones that were never designed to be routed such as *NetBios Extended User Interface* (NetBeui) or packets bearing a nonroutable, local IP address.

At the corporate Internet interface, your packet is processed by a VPN server, which reverses the process. The payload data is stripped out, de-encrypted to its original form, and placed onto the corporate intranet. This process works the same for data coming from the corporation to your remote PC. When it arrives at your PC's Internet interface, it is diverted to the VPN adapter's virtual address. There again, the encrypted data is extracted from the IP packet envelope, converted, and handed off to the PC's operating system. This communication technique is generally referred to as *tunneling*.

Since this interface is internal and software based, your choice of hardware NIC is irrelevant. Wired or wireless network cards use VPN in the same way. Data sent over a wireless interface is considered to be secured from antenna snoops if it has been encrypted by a VPN first.

If you are operating a small business from your home, this technology may be useful even if you are not an employee of a large corporation. It is possible to take wireless throughput from a Wi-Fi access point and feed it into your network through a VPN-enabled router, for greater security. You may also use VPN to secure customer data, such as accounting figures, so that clients can send it safely to your in-house SOHO web server.

Your Own Web Server

Web pages in HTML format have become a universal method for information interchange on the Internet and on local intranets, too. It is possible to have your very own web server as well. Even though the upload speed of a typical residence DSL line is slower than the download link, you can still use yours to provide large amounts of information to the cyberworld, and the PC you choose doesn't have to be very fast. We do not recommend trying this for a home business, however, unless you are ready to spend enough money to do it right.

Not for Business Use You have the option of hiring web hosts who can provide you with services that would be hard to duplicate at home. For example, it is not easy to set up a home-brew counter that keeps a running total of site visits, or to compile statistics about your visitors. Handling online credit card purchases from a home server is no job for amateurs, and you don't want to depend on a residential DSL link for your income. It is true that any money you spend on your SOHO web site, wherever or however you build it, is a legitimate business expense for tax purposes. But servers at home are not as fast or reliable as the commercial variety. If your DSL or cable line goes down, it may stay down for days. A home web server can be a long-term time commitment. You will be spending hours to back your server up, to design and change and maintain it. None of these activities contribute to your bottom line. Here are some other cautions:

- Your ISP may forbid web servers, unless you pay more for a business-level connection. They may even disable DSL interface ports to prevent it.
- Alternatives, including free web space, are available (if you don't mind the banner ads).
- You will need a permanent IP address so the world can reliably find your web server.
- Your router must be configured to accommodate a web server (see Chapter 4, Special Application Configurations).
- Outsiders are invited, so servers are more likely to pick up and spread viruses.
- Web server software costs more than workstation software.

All that having been said, if you enjoy a technical challenge and you are willing to tackle this project strictly for fun, then a home web server has advantages.

- A local web server can be a backup server or other Wi-Fi mass storage.
- You can use it as a workstation.
- Unlike rented facilities, you can experiment with hardware and software.
- Aside from your labor (which can be considerable) no monthly charges.
- You can set up an FTP server to download files on demand.
- You can make your website look however you want.
- Building local web pages is easier when you don't have to upload them.

- Web pages for friends, relatives, and special events. Lengthy audio-visual files.
- It's easier to set up local Webcams, Internet weather stations, etc.

A home web server can be a good training exercise for the acquisition of useful job skills. As the world turns to web pages as a common medium of exchange, it can't hurt to know the universal language they employ.

If you lack a static IP address, one workaround is to put a button on your permanent, ISP-provided home page that redirects users to the IP address of your cable/DSL modem. This in turn sends them through your router's DMZ to your web server. When your IP address changes, you will have to change the button address to match. As we said, it's fine for hobbyists and experimenters or those who want to test a web page under development by putting it online.

All the activities in this chapter can be done today, but the field of personal communications is young, vibrant, and growing in unexpected directions. Most "new" inventions are actually new combinations of existing technologies. As the computer, the camera, the radio, and the Internet merge, they are creating tantalizing possibilities for the world that awaits us. In our last chapter, we will ask some experts and one novice to peer into the foggy but exciting future to see what we can expect.

Where No One Has Roamed Before

She is the latest in technology, almost mythology, but she has a heart of stone. she has an IQ of one thousand and one, she has a jumpsuit on, and she's also a telphone.

—Yours Truly, 2095 by Electric Light Orchestra

The future—in this chapter, we ask knowledgeable players in the *wireless industry* (Wi-Fi) industry to tell us what we need to know about the medium today and to give us their views of what lies ahead.

Boingo Wireless

Christian Gunning is the director of product management for Boingo Wireless. Boingo provides Internet access and other services at hundreds of hotspots across the country. He spoke from his office in Santa Monica, California.

Do you think the day will come when Wi-Fi users can connect from anyplace on the planet?

It's probably not Wi-Fi itself that becomes ubiquitous. There are still significant portions of the world that don't have access to a simple telephone line, right? The future is more of a collection of different wireless technologies that you seamlessly move in and out of. In the future, the Internet will be accessi-

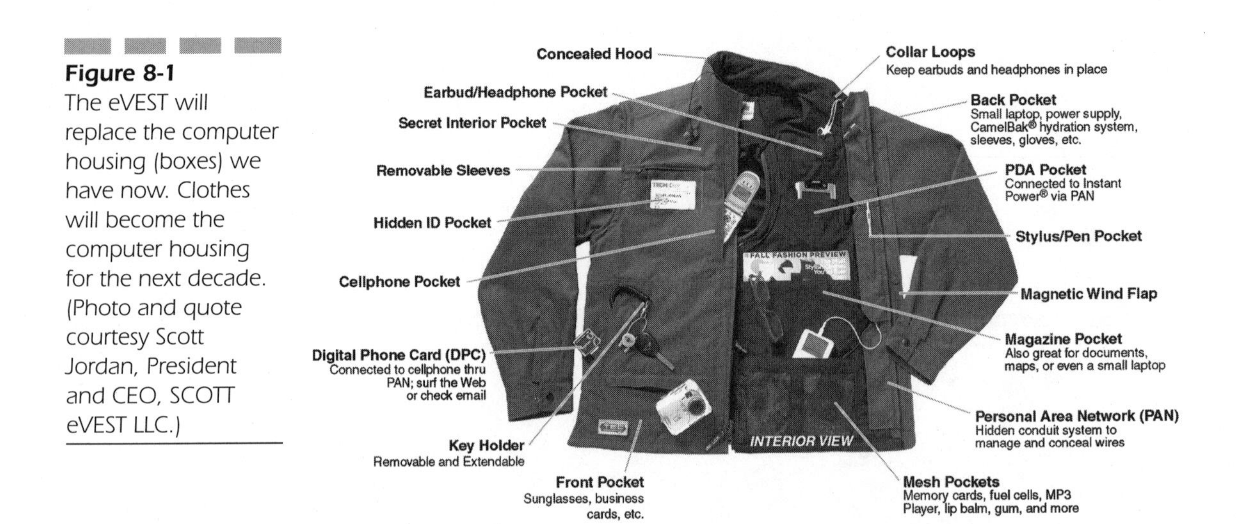

Figure 8-1 The eVEST will replace the computer housing (boxes) we have now. Clothes will become the computer housing for the next decade. (Photo and quote courtesy Scott Jordan, President and CEO, SCOTT eVEST LLC.)

ble to you as you travel the course of your daily business across the country and throughout the world.

What we see in the near future is that Wi-Fi will be used heavily in hotspots and in home networking. You will probably see a proliferation of 802.11a in the home as people start to move toward networked media centers. You'll start to see higher bandwidth in the home. And you will see high-bandwidth along the travel ribbon for business travelers and in static high-loiter locations, where they are sitting for a while. Outside of that, you will have 3G connectivity, which is a medium-bandwidth solution that will keep you connected while you are moving. At some point, does it start to incorporate satellite technology or other things? It's all possible.

Are users going to accept this new technology?

What we're starting to see is that people are getting more accustomed to having an Internet connection. They are having epiphanies where they realize they'd really like an Internet connection *right now* to check on *X* or to do *Y*. Presently, since it's largely a wired world, those things are not all that easy to do. As we move toward a wireless world, you see the foundations being put in place. With public hotspots and the rollout of 2.5G and 3G (cell phone Internet) technologies, that availability is going to be there.

Why not now?

There are challenges. Number one, some of the hardware prices have to come down. Wireless networked PDAs are still sitting at the $500 to $600 mark. That's not the right point of inflection for customer adoption. It's got to get down to the $200 or $300 range before people begin to adopt them en mass. And a behavior change is also necessary. People will have to look beyond their cell phones and PDAs as phones and rolodexes, and start thinking of them as network devices, a point of access to get them everywhere else they want to be. We're starting to see multimode devices and that is going to accelerate. They will be intelligent and realize that there is a faster or wireless connection available, or switch to 3G mode if moving 60 miles an hour in a car.

As users switch channels and move about, who will keep track of their minutes and bills? Will Boingo set up partnerships with large phone companies and other hotspot providers?

That is exactly what we are working toward. We talk to every wireless ISP that's in the process of deploying locations. We're also talking to cellular carriers about reselling their 3G services. We see our service as band agnostic. In the future it won't be a question as to which of these technologies will win. They are both complimentary and they both do different things better. We're trying to build intelligence into our software so that it works in multi-, or dual, or tri-mode network cards, so we can move from network to network. We are partnering with companies like TSI, who are roaming providers for the cellular industry, to help us combine the cellular and Wi-Fi worlds. We want to make it easier for the carriers to do joint authentication and billing.

Some providers, such as NYCwireless, are passing out Internet access for free. Does that seem viable to you?

A lot of enthusiasts across the country are doing free wireless access. It's more of a hobby initiative. The reality is that Wi-Fi is a short-range radio and has limitations. Our business right now is focused primarily on business travelers. As long as that's the case, there are primary properties, airports and hotels, and places where business people are spending a lot of time, in which the owners of the real estate at those locations will have control over whether or not Wi-Fi is accessible. Lighting up a public park is great for the community. It gives people a place to go to get connected. But over the long term, who is going to provide support for the people who can't get connected? Free community networks are not going to put up an 800 number and staff it around the clock. Business users need reliable connections and are willing to pay for someone to help them troubleshoot connectivity issues, or make sure that they actually get connected, and get their mail done, and get whatever corporate access they need.

Could you say the same things about ad hoc commercial providers, like the small-time hotspot providers of Joltage.com?

They are really entrepreneurial-driven wireless ISPs. They download software and configure their machines to become a local access point. Then it's a commercial location that they actually charge for access to. I don't know what their customer support perspective is or how they deal with outages, whether they have a page that shows network status. I see them more on the commercial than the free side.

What's the most important thing we should understand about this technology?

The biggest thing is that we're going to reach a point where the Internet is just in the air we breathe, right? We're not going to think about getting on the Internet. It's just going to become an essential part of the fabric of our lives. It will be second nature. A few years ago cell phones were not all that prevalent, and now almost everyone has one and everyone doesn't think twice about getting on that phone. You look back before that, and long distance was expensive and it was special to make a long distance call to family. Now it's gotten to the point where everyone can pick up the phone and call long distance whenever they want.

You are going to see that same kind of adoption and embrace with the Internet. In the next few years, it's just going to be there and people are going to start using it without thinking twice.

NYCwireless

Anthony M. Townsend is the cofounder of NYCwireless, a public access ISP in New York City. He is also a research scientist at the Taub Urban Research Center at the Robert F. Wagner Graduate School of Public Service of New York University. In addition, Mr. Townsend is the President of Cloud Networks, Inc., a startup firm that develops wireless infrastructure for commercial properties.

What will be the effect of Wi-Fi technology on the way we live?

People won't notice it. They'll notice when it is absent. Places that don't have an Internet connection will become marginalized. They will be viewed as second-class places in the same way that locations without electricity or air conditioning are seen today.

Mobile wireless communications are making our lives move along faster. You can change your plans en route or change other people's plans en route. It gets rid of the wasted time that we used to spend sitting around waiting for someone to call or for mail to arrive. We can be much more tightly coordinated because of that.

When the telephone was first introduced to the home, it was often regarded as a tool for work that should have stayed at work, sometimes as a threat to home life. Will ubiquitous Wi-Fi be seen the same way?

If anything, it will allow you to be more flexible about where you locate your home office. It won't necessarily spread work throughout your home. Even if it does that, in the United States in this day and age we are not greatly opposed. In fact, we are welcoming it.

I think the more interesting way in which those boundaries are being broken down is between office buildings and parks. You can do serious work in a park with a cell phone and a laptop, if it has wireless. We're seeing that in Manhattan in large office districts. People use their VPNs to connect to work and stay in the park for four hours until their battery dies. It's really amazing.

Isn't there a downside to that? Computers were supposed to liberate us from drudgery, but you see people using them in restaurants and on airplanes. If people can work anyplace and any time, won't their employers demand that they work anyplace and any time?

The idea of machines doing all the work for us is in the same overly simplistic category as the paperless office. Before the industrial revolution, home and work and social life were not separated in time or place. Before the steam engine, most folks had a workshop or production capability inside the home. The home was often just one room, where all these activities went on together.

That's the way things were throughout the history of humanity. It became separated during the industrial revolution for a variety of reasons, but mostly because the production apparatus was so gigantic—physically large—that it called for a separation of uses.

As we get these technologies that allow us to be more flexible about the workplace, we are very, very quickly drifting back to a preindustrial approach. This is an emerging line of thought in urban planning too. Information technologies can let us reestablish cottage industries in a way that we haven't seen in a long time. It may be different from what we've had for the last 50 or hundred years, but it's probably more natural to have some work at home. Wireless networks can support that and make it more pleasant. Maybe you can work while sitting on the back porch instead of sitting at a desk. Or sit at the kitchen table and take care of your kids instead of sitting in an office.

Is your view of "dead spots" as second-class spaces the motivation behind your effort to "light up" New York City parks?

Absolutely. We wanted to make them just as good as corporate or university spaces. We also wanted to claim them as free wireless spaces before anyone could step in and set up a pay network there.

Some commercial providers think that without a profit motive to sustain them, freenets will eventually have to give way to pay nets.

Naturally, I don't agree with that. I think there are some flaws in their conceptual thinking. The World Wide Web went through a noncommercial phase, then a phase with lots of commercial activity, and now it's starting to revert to its pre-late-1990s state, where it's disorganized and search engines don't index everything. But it's still working fine.

The two models for spreading wireless are basically the wireless carriers model and the fiber-optic structure model. The carrier model will probably not happen because they have so much debt, and they are having trouble making money off the services they already provide. The other model is real estate based. People will see the value that wireless provides to their building or neighborhood or home, and so they will invest the money to set it up themselves. I'm convinced that's the way it will go because that's how most of the last-mile fiber was built out. It was actually put in by electrical contractors rather than companies such as Quest or Level 3.

A lot of companies like Boingo are trying to create a quick and easy solution to creating wireless hotspots. There isn't one. It's very piecemeal, step by step, bit by bit, which is why it lends itself to a grassroots movement rather than a top-down approach. NYCwireless is already one of the largest ISPs in North America. We have 80-plus hotspots. That's more than Starbucks, which is the second largest in New York City.

Are there other organizations like yours in other cities?

We all work together loosely under the banner freenets.org, which is also the URL of their web site. Groups in different cities are proficient at different things. The ones in California tend to do a lot of software work. We tend to do a lot more of the organizing, public relations, documentation, and we've looked into the legal issues more deeply. We complement each other's efforts.

What is the most important thing that people should know about Wi-Fi?

This is going to be a negative thing. The most important thing is security. Wireless networks are inherently insecure. It's like walking into a public place and letting a complete stranger plug into your laptop. Everyone picks up everyone else's traffic. It's easy to search though that for passwords, credit card numbers, and things like that. The most important thing is to always use N10 security, SSL, SSH, VPN, whatever.

We are finding that security is the single greatest obstacle to the future adoption of this technology. When the security risks become clear to the everyday user, they will probably stay away from it. It makes what's happened with security on the Web look like a walk in the park.

Linksys

Jim Harrington is the product manager for the wireless division of Linksys, a leading manufacturer of broadband networking equipment for the home and office.

How will Wi-Fi technology change our lives, particularly in the home?

I think it will change our lives quite a bit. From the first time I picked it up, I said "Wow! This stuff is really going to take off!" When I first started toying with it, I was hooked immediately, and that seems to be the effect it has on most consumers. The tricky part is getting them to understand this new technology when networking in general is tough to pass on, especially to home networking users. Customer education is always a concern. We're trying to make it as easy as possible, so as long as you're using Linksys stuff, it'll all work together out of the box. But it's like anything else. It's constantly evolving, with different standards coming out.

Given the rapidly changing technical environment, predicting the engineering future must be difficult.

It is tough, because we rely heavily on the chipmakers, as does everyone else. We try to allow space on our hardware, on the physical layer, to adapt to changes that can be made through firmware. Otherwise, we have to go with whatever the standard is and try to make it backward compatible with legacy products.

Is that always possible? How likely is it that some buyers will get left behind?

If you look back at wired networking, in the beginning there was arcnet and token ring and even different standards for Ethernet. It had to shake out before everybody settled on twisted pair wiring. That's kind of what's happening with wireless. There will be a few different standards. I personally think that 802.11b and g will still have a very strong place in the home networking market. They were first to market, and the cost is significantly less than the 54 Mbps a equipment.

Wi-Fi makes telecommuting easier. Does it mean the end of large corporate offices and their cube farms?

Broadband in general would shape that more than wireless by providing high-speed access to VPNs, with or without wires. Personally, I think you still have to have that human contact and a daily association to make most companies function. As for the virtual office, I don't think that we're there yet. Even networking in the home, well, the other day my wife sent an email from the living room to me upstairs, telling me that it was time for dinner. We laughed about it, but we're the exception. At least, I hope so.

I forget the exact numbers, but something like 15 percent of homeowners today are either networked or considering it. We have a long way to go before broadband is settled down enough so that we can do something like a virtual office.

Some folks think that broadcasting to the neighborhood is okay. Do you think a day will come when you can open a laptop almost anywhere and get to the Internet?

I think that's probably true. Whether public spaces will be free, like NYCwireless and Baywaug and Seattle Wireless are trying to do, well, somebody's going to have to pay for that bandwidth eventually. Firms like Boingo are trying to come up with a solution, which is a little pricey right now. But yes, I see it as a definite possibility.

What engineering solutions will engineers provide for Wi-Fi security?

There are solutions like the 802.1x standard or RADIUS, but that's high end and more appropriate for the corporate level. On an average wireless network, we're waiting for 802.11i, which will overcome some of the weaknesses inherent in the present security standard called WEP. We're doing some things on our end as well, but we want to stay with standards and we don't want to get too proprietary in our solutions. Most good vendors will offer MAC address filtering, disabling SSID, and access point transmit power throttling so that you won't broadcast to your entire neighborhood if you don't need to.

What's the most important thing to know about Wi-Fi?

As much as I hate to admit it, because it has been overhyped so much in the media, it's security. The more wireless proliferates, especially in smaller communities, condominiums, and so on, the more you need to pay attention to the security features—if for no other reason, than just to make it work. My neighbor has a wireless network, and I could see their access point from my system as soon as they hooked it up. I went over there and said, "You need to do something about this."

I hate to scare everybody though, because it takes a lot of savvy and a great deal of knowledge and the right equipment and a lot of patience to hack a wireless network, if the network is using the security tools provided.

The Wi-Fi Alliance

Dennis Eaton is the current chairman of the Wireless Ethernet Compatibility Alliance. In addition, he is a strategic marketing manager for Intersil, a Wi-Fi Alliance member corporation based in Irvine, California. Dennis has over 16 years of experience in *radio frequency* (RF) design and holds degrees from Michigan State University and the University of Central Florida. He actively participates in both the *Institute of Electrical and Electronics Engineers* (IEEE) and the Wi-Fi Alliance. Also participating briefly was *C. Brian Grimm,* the Wi-Fi Alliance's Marketing Director.

The Wi-Fi Alliance checks equipment for adherence to standards, but there are a lot of standards and more on the way . . .

Eaton: Sure. We call them the alphabet soup. Customers should focus more on what's happening in the Wi-Fi Alliance and less on what's happening in the IEEE. That will help with the confusion a little bit. Another thing is that many of the new 802.11 standards are extensions to the physical layers, the radio standards that are already in place. They're not going to break your equipment or make your network obsolete. They are merely new features that are coming down the road.

Some devices using the Bluetooth standard seem to do the same things as Wi-Fi tools. How does a buyer choose between them?

Eaton: There is a little overlap between the two. If the user looks at Bluetooth and thinks "cable replacement technology" and looks at Wi-Fi and thinks "wireless replacement for my Ethernet connection," that will help to clarify things. The two technologies, though they cannot communicate with each other over the airwaves, are very complimentary. They were designed to do two different things: Bluetooth, to get rid of the rat's nest of cables surrounding your computer, your PDA, and your cell phone. Wi-Fi was designed to plug into your existing Ethernet stack to replace your high-speed network connection. There will be some vendors who choose to take a technology in a direction that it wasn't designed for—Bluetooth vendors who try to make a wireless LAN, and maybe vendors who try to force Wi-Fi into an ad hoc serial cable replacement technology. But those are a small minority. If consumers focus on buying the right technology for the right application, then they will be okay.

How will Wi-Fi technology in the home affect our lives? Will it change our social structure?

Eaton: Maybe a little bit. The primary thing that Wi-Fi does today is to give us mobility. It allows us to do the typical things we already do—surfing, email, that sort of thing—but without constraint to a desktop. Wi-Fi makes your connection to the Internet and all it provides more casual. Wi-Fi also provides an easy way to connect multiple computers without running new wires or trying to use the existing wires in their house that weren't designed for data networking.

That's how Wi-Fi is used today, but in the future, we'll probably see it work its way into consumer electronics. It will allow you to buy a TV receiver that's Wi-Fi enabled and very easily stream video channels to a monitor or speakers that are also Wi-Fi enabled. Or connect with different digital devices in your home, such as an MP3 player. Someday you might have wireless telematics to your car, so you can send and receive MP3 files to your car stereo. Those things are out there, probably in three years.

Will home Wi-Fi make it less necessary to go in to the office to work?

Eaton: There's a little bit of truth to that. Here where I work we've been using wireless. We have more people work at home on occasion than we would otherwise. Some of that is because Wi-Fi does make it easy to connect to the network at home. The other part is that Wi-Fi very much goes hand in hand with the rollout of broadband to the home. Once you get many homes enabled with broadband, be it DSL or cable, and you marry that with Wi-Fi to make it easy and casual to connect, then that does knock down some barriers. I don't think we'll ever completely get away from the need to come in to the office.

One of the things that Wi-Fi does is change the office work dynamic in a good way as well. People can remain productive and tied into their desktop applications while they go to meetings and collaborate with others. You see an increase in productivity because employees don't have to set aside work. You see more collaborative and group work.

Will the day come when people can use laptops to connect to the Internet outside of hotspots?

Eaton: Wi-Fi is very complimentary with what we are told that 3G (advanced cell phone) technology will be. Wi-Fi provides a much higher rate connection than 3G in a very targeted way. Very specific locations where people collect and remain stationary for short periods of time—airports, cafes, and hotels—these are a very good application for Wi-Fi. It allows a provider to roll out very high-speed data services in a very economical way. There's no licensing fee to use the frequency, for example.

One of the things that 3G technology can give that Wi-Fi cannot is ubiquity, or coverage. 3G can take over as you move from the LAN to the WAN, and provide connectivity to your application, though perhaps not at the same data rate you experience with Wi-Fi. I think you'll see a lot of things happening in the next year or two as people bring the two technologies together in a single device or network connection, and roam between the two as well.

What's the most important advice you would give to a new user about Wi-Fi?

Eaton: If they are going to use wireless, they should focus on Wi-Fi because it's interchangeable. They can take their laptop to the office or another Wi-Fi house and know that it'll work there. But the mobility that it provides can also make you more susceptible to allowing others to get on your network or see your network. It's great that Wi-Fi can flood your home with RF that allows you take your high-speed connection with you, but the RF doesn't stop at your front door. Somebody passing close by could also use it to get connected, if you do not protect yourself. You need to enable the security features that are built into the equipment. That's something that many people don't do because it is so easy to set up and use.

Also, and I say this kind of tongue in cheek, we have cell phones and pagers and it's all good stuff. It allows us to stay connected and be more productive and that is a good thing. But it can also be a bad thing . . .

In what way?

Grimm: I can give you a good example. You can lose your love life! (General laughter.) I'm serious! I have a wireless laptop, and my wife Donna has a wireless laptop. She's on eBay until 11 P.M. In bed. Outside of the office, we use the Wi-Fi connection more in bed than anywhere else.

Eaton: Yes, you can literally take it anywhere with you. Like the cell phone, Wi-Fi allows people on the Internet to reach out and touch you no matter where you are. It makes you more accessible. That has its upside and downside.

In-Stat/MDR

Gemma Paulo has been an analyst with In-Stat/MDR for the past 3½ years, specializing in *wireless LANs* (WLANs). In-Stat/MDR is a commercial research and analysis agency serving technology vendors, service providers, technology professionals, and market specialists worldwide. Gemma spoke with us from her Scottsdale, Arizona, office after recently completing a business trip to Korea.

How will home wireless change us as a people?

That's a good question, one that everybody wants to know the answer to. Wi-Fi has been interesting because we did not think that Wi-Fi in it's present form, 11 Mbps if you're lucky, would be such a force in the home. The numbers are really high, considering that most people are networking just one PC, sometimes two or three, to their cable or DSL modem. Often users are unconcerned with the lack of high bandwidth and are just happy to be able to get remedial Internet access. WLAN has been hot in North America as well as Japan and Korea. The equipment is very, very cheap. We had no idea that prices would go down so far. There are close to a hundred vendors in the market providing cards and access points. Never have we seen a market that was invaded by low-end, low-cost vendors so quickly—within a year and a half. It's just been incredible.

What makes it so popular?

The 802.11b products have been on the shelves since late 1999. The low-cost vendors who sell retail are on their third generation of product. The software is better now. Judging from the few that I've tried, configuration is very easy. The cables are color coded. You stick a CD into the drive, it pops a client into your laptop, the NIC finds an access point for you, and your network will be pretty much set up. Microsoft recently announced their intention to get into the market claiming that their solution is going to be the easiest out there. Wi-Fi is widely available at retail, and also on e-tail. People like to compare prices online.

How is Wi-Fi evolving?

We are seeing more complicated multimedia applications such as streaming audio and video, say, to and from your TV. Everyone is looking toward 802.11a or 802.11g to serve that function. The one thing that's missing is quality of service for multimedia streaming support. A lot of people have suggested a complimentary technology to solve that, such as ultrawideband, and others look to the 802.11 Working Group to figure it out. The multimedia issue will be overcome. That's what everybody is waiting for.

Another question is, what form will the residential gateway take? Some kind of set-top-box, combination cable modem? Streaming from some kind of central media server to a variety of devices, such as MP3 players, camcorders, DVRs, and televisions? We thought that consumer electronics companies would be in the market a little faster than they actually have been. We haven't seen a lot of embedded Wi-Fi in these devices yet, and we probably won't see it in large volume for at least two or three years.

If you put the benefits of mobility aside, is there a downside to this technology?

Many people have separate home versus business networks. Usually, you don't cross them. But if you do telecommute—there is lot more of that now—or if you do any kind of work from home, then Wi-Fi is not the safest thing. Users don't turn on web security options that frequently.

Another problem has to do with roaming from hotspot to hotspot. It is almost impossible to stay on one subscription plan. You have to sign up for several. Often mobile workers discover that Wi-Fi won't work at their hotel, or that there are problems setting it up. I would say that roaming is not really easy to do yet.

One problem for the cable or DSL guys is shared Internet access. Someone who pays $40 per month for a residential cable modem can share it with a lot of people using a wireless LAN. Providers really don't want that freeloading to happen.

What can they do about it?

One vendor I know of wrote malicious letters to many of their customers and made some of them angry. Wireless home LANs represent a small part, perhaps 5 percent, of the broadband total. But it is in its infancy.

How big is it? How big will it get?

I'll look that up. All right, in 2001, there were about 2.6 million units shipped, access points and clients. By the end of 2002, we expect about 5.6 million additional units to ship out. It's not that many worldwide, but it's a fast-growing market compared to the others right now. The producers that have done well are traditional vertical players who sell to people who need to be mobile, or networking vendors who know that they need to have a wireless LAN solution to get into an account so they can sell their higher-margin switches and routers. There are also low-cost vendors who are quite used to selling high volume, low margin. Wi-Fi is not their highest profit maker, but, one, in a sense it's kind of exciting and, two, I would guess that it gets them into higher-margin accounts.

Home wireless is a tough market to be in. You have to deal high volume and low cost. In our research, we have tracked many different technologies, such as HomeRF, phone line, and HPNA. In terms of seeing the number of vendors and solutions, this technology is very interesting. It seems like it has a big future.

What's the most important idea about Wi-Fi home networks?

Just how odd it is, working with a radio frequency that goes through walls. There are no real rules. The type of wall might screw it up or the number of floors in your house. The manner in which you place your antenna might increase your signal two or three times. Sometimes you might need a little guidance from someone who has installed one already, or look on the Web for pointers. I would say that it is an exotic technology.

Apple Computer

Dave Russell is Apple's director of worldwide product marketing for mobile consumer and wireless products. Dave has been with Apple for 15 years. He was instrumental in launching their wireless AirPort in the first iBook several years ago.

When telephones were first installed in homes, they were sometimes viewed as a threat to family life. Do you think home Wi-Fi might be perceived this way?

From my early days, I remember that phones were big, clunky, rotary-dial things that stuck out from the wall like an afterthought. They didn't look as though they really belonged. They looked more like a necessary evil. If you want to draw that analogy forward into home networking, you can fast forward from the phone as a business tool to something that's a ubiquitous consumer product, something that everybody not only wants, but also virtually has to have . . . access to other people or businesses, or access by other people to reach you in emergencies, or just for fun—that's an assumed part of life in the modern world. I think wireless networking in the home allows those same benefits.

Wireless networking allows every member of a family to use their computers, even simultaneously, and use them anywhere, comfortably. Prior to wireless networking in the home, if you wanted access to the outside world it usually meant a dial-up, and later, cable or DSL modems. Regardless of the throughput differences in those technologies, the computer always had to be located close to the modem or phone jack. That's not necessarily the best place for computing. By putting an AirPort—that's our trade name for Wi-Fi technology—where the cable, DSL, or dial-up connection comes into the home, you can serve any room in the home. The connection radius averages 150 feet, and most homes are smaller than that.

Why is that important?

Because of the places in which adults feel comfortable doing what they want to use their computer for, say, managing finances, doing research or work, and sending it back to the office. They may want to work in a business-like, quiet environment, such as a den. A schoolchild trying to do research or collaborative learning with another student probably wants to do homework in the kid's bedroom. There they can express themselves and feel capable of working at their own leisure, without the distractions of parents or that oversight. Wireless networking allows that and more. If you have a portable computer, you can roam about the house and on an ad hoc basis be connected to the Internet and get your email wherever you feel comfortable. That applies outdoors as well because you can take your computer outside. I think that's one of the reasons that wireless has caught on so dramatically.

The majority of the portables that Apple sells are purchased with an optional AirPort card in them or with one built into the machine. So we know the technology has hit home—no pun intended.

Another benefit, especially in a dial-up situation, is that one base station, using DHCP, can share an Internet connection using one ISP with up to 50 users. You don't have to buy a relationship with your ISP for every member of the household. The expense goes down because of that. And before, if you had multiple computers, you'd have to argue over who was going to use the single connection at that moment.

Users with broadband connections such as cable or DSL modems in the house get high bandwidth. There is a perceived downside that comes with that —the potential for security breaches into the home by virtue of your computer being connected all the time. If you use an AirPort base station to connect your broadband link, it has a number of cascading security features to prevent hacks and attacks. The station has a built-in firewall so the outside world cannot see the Ethernet ID of the home computers that are connecting. 128-bit encryption is a fundamental part of the base station. We offer a lot of flexibility in terms of being able to name your network, secure that with a password, and close that network such that unless you already know the name you could not find it.

It sounds as though Apple is firmly committed to this technology, going forward.

We are. We like to think that we revolutionized that technology. We didn't invent it, but we saw the obvious benefits of it in the late 1990s. We saw then that it was also very expensive, not easy to use, and at 1 or 2 Mbps, not very fast. It had been adopted by only a few vertical industries for things like inventory control or handheld devices. Wireless NICs ranged from $300 to $600 per card. Base stations—we call them airports now—were selling for $1,200 to $2,500 each. And the stuff was hard to use. You had to do a lot of networking configuration and setup. It was anything but a consumer technology.

Apple popularized it by bringing the cost of a wireless card down to $99 and the price of an access point down to $299. That's 20 percent of the industry standard pricing at the time. We built very nice, easy-to-use AirPort Setup Assistant software into every Mac. From mid-1999 forward, with the introduction of the iBook, every new Macintosh came with built-in antennas and a dedicated slot to take an AirPort card. We embraced the technology, pushed it forward, and popularized something that was in a chicken-and-egg-type situation. We're sort of famous for doing that. We did it with USB and Fire Wire, and today, with Bluetooth.

Will a non-Apple laptop work with an Apple AirPort? How easily?

One of the benefits of Wi-Fi today is that it's based on an industry standard called 802.11. They are interoperable. A Macintosh notebook or a Wintel laptop can go to a Starbucks or an Ambassador's Club at an airport and get on to their networks. They'll work in a complimentary fashion.

Where do we go from here? Perhaps a do-everything wireless box that enables you to control your home's air conditioning over the Internet? Streaming video?

In general, what I think you are going to see in the short term is much more ubiquitous implementations of wireless capabilities. We're beginning to see it in airports. (I don't mean our brand name, *AirPort,* but the ones where airplanes take off and land). Starbucks is implementing it with a company called T-Mobile in thousands of coffee shops around the country. The capability to wirelessly connect will become much more prevalent.

Beyond that, I think there are the capabilities to do the things that you suggested. The technology would accommodate that. The future is very bright for this technology. It will enable a lot of other devices that previously did not connect well to connect wirelessly. That will be a tremendous benefit to some of those devices. The fact that the technology is standards based and interoperable regardless of whether it's a Macintosh, a Windows machine, or an appliance allows many other wireless devices to be created.

Wireless makes it easier to work from home. Will this bring an end to the corporate cube farm?

I wish I knew, because I'd like to work more from home. Wireless allows you to maintain your connection to work from anywhere in your home, which is pretty handy. To decentralize the workplace out of the corporate environment into the home is a social issue that's probably bigger than just wireless technology, but wireless could enable that to happen. There are many other things to consider, the value of personal interaction in terms of the idea creation that happens by having people actually rub elbows and see each other face to face. My sense is that it's possible because of wireless, but I'm not a social scientist. I'm a technology guy. So I'd love to believe it.

What's the most important thing for a wireless newcomer to know?

Buy a product that's interoperable, because they'll want to take advantage of public wireless facilities. Ease of use, and having your computer designed from the ground up to take advantage of wireless is important. For example, you'd prefer a laptop with a built-in antenna because an antenna sticking out of its side is fragile. It can break off, or someone could pull your card out of your PC and walk off with it.

If I can reiterate, Wi-Fi provides people with the capability to comfortably access the Internet and do email wherever they want to, in their home or in a public place. Given that, it has every potential to increase personal communication, which is a wonderful thing. It allows families to maintain much closer ties. It's easier, certainly faster, to communicate by email than to send letters through the postal service. Not to mention that you can attach photos and other digital imagery to enhance it.

Beyond that, the ubiquitousness of this technology increases access to information on a personal level. As an example, the ability to quickly find a map to a person's house is tremendously useful to me—to be able to find a recipe for dinner, or look up information of general interest. The other day I needed to know how many moons were around Jupiter so I just got on the Internet and looked it up. There are 17, plus a ring system. Things like that are hugely beneficial to everybody because this technology allows us to compute anywhere and at any time.

Beyond that, it is truly a boon to the learning experience. U.S. educational institutions, primarily K-12 and colleges, have been great adopters of wireless technology. It gives Internet and computer access to every student without having to worry about them tripping over Ethernet wires.

It must be exciting for you to be playing a major part in the evolution of it.

It is really exciting. I enjoy my job. I enjoy looking at capabilities that haven't yet become popular, and making them popular, and seeing how they can change people's lives. It's actually very rewarding.

Hewlett-Packard

David Chapin is HP's home and small office product marketing manager who has been with the company for two and half years, following the merger with Compaq computers. Prior to that he worked with *Computer Aided Design* (CAD) in manufacturing software, dealing particularly with distributed processing across networks.

Why did you get involved with Wi-Fi?

It's kind of exciting and rare to find these days in the computer industry. Wi-Fi is a growing marketplace. We have the ability to take all these great end-user, access devices such as MP3 players, digital jukeboxes, PCs, notebooks, handhelds, and bring them together. That's kind of neat. We see a lot of different applications and challenges to go along with that.

Describe HP's future participation in the home Wi-Fi market. Is it here to stay?

Absolutely. If you look at where HP has taken wireless just in the last couple of years, we've driven it down into consumer-level products like notebooks and wireless handhelds, and in a wide range of wireless LAN, WAN, and PAN solutions. It's certainly being integrated in the bread-and-butter products of HP. Groups like mine are working on solving the wired and wireless problems of network neophytes so they can get our products pulled together without needing an MIS degree.

How will standards evolve? Do you think that home network setup will ever be completely effortless?

The evolution will follow the same path as technology in general. The first computers presented you with the DOS operating system, and you had to be dedicated in order to deal with it. Now we're at Windows XP, where you boot the thing and many of the features and functions that you use a PC for are already there for you. Networking and connectivity will take that same sort of trek, and it will all be based around standards.

A lot can be said about Microsoft on both sides of the fence, but one thing they did was provide a singular platform across many different manufacturers' hardware. The IEEEs and Wi-Fis of the world are helping us do that, from standards definition up through a standardization of marketing. Wi-Fi plays a valuable part in just trying to get the word out and allowing us to wrap our solutions around something easier to understand than 802.11 and its variants.

We've heard predictions about gadgets like universal home gateways and Wi-Fi-enabled cars. Where can we go with this technology?

I think Wi-Fi is always going to live as a local area network. It will certainly have its place in the world of connectivity between your WANs and PANs to give seamless connectivity anywhere anytime. If I put on my dream cap, I think it would be very possible for a central brain device that would bring all of these networking media—possibly even different types of networks—together, although I do think that IP networks have become the de facto standard. Then we need to get the right network media, be that phone line or power line or wireless Ethernet, whatever makes the most sense in a particular environment. Residential gateways can play a crucial role in bringing all that technology together, so you don't have to worry about media.

The remote manageability of these networks has proven out in the enterprise space, so it should be possible to do that in the home. Things like IPv6 could possibly do away with NAT to make it easier to find your home in the world, rather than always having to deal with dynamic IP addresses.

While waiting at a car wash, we noticed that most of the customers were talking on their cell phones instead of to each other. The thought occurred that when the car wash owner puts up his hotspot, perhaps the customers will bury themselves in their laptops and stop talking entirely. (Chapin laughs.) *What is this technology going to do to us?*

I would say that the proverbial glass is half full there. They were communicating with somebody, perhaps not the guy standing next to them, but someone important to them was on the other end of that cell phone. So, if anything, we will probably be better connected. From a personal interaction standpoint, you could argue that there are downsides, but it's not as though connectivity technology is isolating us. They were talking, not playing Game Boys.

Easy Internet in the home means easy teleconferencing from there. Workers won't necessarily have to commute to work.

I think that's a nice vision, but I'm not sure that it works in every environment. If your entire day is spent in front of a computer, and you did not need personal interaction, then sure, why not? But there's still a lot of creativity that goes on at work, and a lot of it has to do with teams of people interacting with each other. You can videoconference, but there's still nothing like the personal touch. I do have something of a sales background. I like to be there and see the look in their eyes and feel the grip of their handshake when important decisions are made. That's my own opinion. I've also worked out of my home before, and it was quite nice. It allowed the company to hire me for a remote location. They could not have afforded to lease office space to put me there otherwise.

Will we see the day when people can connect to the Internet from anywhere?

Absolutely. It's a matter of infrastructure and data compression to achieve the speed we need at a not-so-painful rate. Think about it. Three or four years ago wireless networking was a foreign concept. Today you can walk into any of the Admiral Clubs and the like at airports, and you have access. We are already getting that public hotspot infrastructure rolled out. It's quite successful. Starbucks is installing hotspots in their shops and HP is playing a big role in that effort. From the office to the coffee shop to the airport, and you have access to the information you need to do your job.

What's the most important thing to know about Wi-Fi?

If I didn't understand it, the one thing that I would be concerned about would be my personal security. Wi-Fi is *reasonably* secure for the home, but not perfect. It takes a lot of decisions out of network topology because all you need to figure out is where you want to put your access point. You don't need to worry about jacks and drops and wires and other wired problems.

Any other comments?

With the present growth of this technology, and the bright future that wireless IP networks have, especially local area networks, the cost of these things is going to come down tremendously. The integration into devices that people are already carrying around will be very affordable. In the end it will make life easier because we already have digital devices we want to connect and it's still kind of a pain in the butt to do that. If you have a handheld, you have to buy a sleeve and a card and crawl over the WEP encryption parameters. As we work this down, you'll have very secure, integrated wireless devices that will really solve some problems for people.

It sounds like it must be rewarding to be involved in this technology.

It's an opportunity to help write history. I do believe that. I can look at the adoption rates for broadband in the home and see some of the things that are now available online at phenomenally reasonable prices. I can imagine that in

5 or 10 years we will have home access to the world's library for $1.95 per month, or access to all the recorded music in the history of the world for $2 per month. It's awesome. It will be at our fingertips and really be possible because somebody's data plumbing is back there tying it all together. We're working to make it HP's plumbing.

It's just a means to an end. We need to make that as easy and as secure as possible, and then get out of the way so that people can do what they need to do.

The Last Word: The Author Gazes into His Crystal Ball

The 1967 film *The President's Analyst* was a political satire about a psychiatrist who quit his job as the president's shrink. Spies from all nations tried to kidnap Dr. Sidney Schaefer, played by James Coburn, to learn the president's darkest secrets. The chief plotter turned out to be a secret organization known as TPC. In the last reel it was revealed that the acronym stood for *The Phone Company*. This fictional organization's grand scheme was to put (wireless) computer chips into everyone's head, as explained to Schaefer by TPC's robot leader:

TPC president: Can you image the ease, the fun with which you can place a call? Why, all you have to do is think the number of the person you wish to speak with and you are in instant communication. Anywhere in the world!

Schaefer: You're a megalomaniac. And the phone company is psychotic.

TPC president: We realize the public has a misguided resistance to numbers. For example, digit dialing.

Schaefer: They're resisting depersonalization!

TPC president: And so Congress will have to pass a law substituting personal numbers for names as the only legal identification, and requiring a prenatal insertion of the Cerebrum Communicator. A communication tax could be levied and paid directly to the phone company.

Schaefer: It'll never happen.

In the end, the film's heroes found the plan so repugnant that they blew up TPC's network control center. But that leaves us with the real-life question: *Will it happen?*

At the end of every chapter in this book, we provided an introduction to the chapter ahead. This is the last chapter, so what follows will be written more by the readers than the author. Wi-Fi home networking is a continuation of a trend that began in the 1960s with the advent of the pocket pager and continued through the cell phone, the wireless laptop, and the wireless PDA. Past today, you can look forward to a combination of all those: a wireless two-way sight and sound communicator that works for anybody, anywhere, anytime. It will be a miraculous and powerful gift for millions. It is probably too late, however, to ask what the customers will have to give up in return. Every new technology has unintended consequences.

As an example, cellular telephone companies are required to install technology that pinpoints the location of users. The intent was to allow those calling for help to get it even if they cannot speak, as we have now with the wired 911 systems. Since the companies will soon have your location anyway, they requested an advance approval from the FCC to sell that information—as a service to customers, they say. That way, users could get advisories on their cell phones directing them to the nearest store, gas station, or whatever. The FCC approved. Soon lots of people you never met will know exactly where you are.

The image that privacy alarmists call up most frequently is the novel *1984*, a dismal world in which every citizen is under the relentless gaze of a television camera. It seemed outlandish in the innocent year of 1948.

Orwell's prediction of the future was inaccurate for several reasons. For one thing, in our twenty-first century, cameras *can* see in the dark. Governments routinely use surveillance to control the behavior of citizens. For

> The telescreen received and transmitted simultaneously. Any sound that Winston made, above the level of a very low whisper, would be picked up by it; moreover, so long as he remained within the field of vision which the metal plaque commanded, he could be seen as well as heard. There was of course no way of knowing whether you were being watched at any given moment. How often, or on what system, the Thought Police plugged in on any individual wire was guesswork. It was even conceivable that they watched everybody all the time. But at any rate they could plug in your wire whenever they wanted to. You had to live—did live, from habit that became instinct—in the assumption that every sound you made was overheard, and except in darkness, every movement scrutinized.
>
> — George Orwell, *1984*

example, many communities across America are placing automated cameras on busy intersections to photograph and fine those who run red lights, in a practice described by one critic as "an Orwellian cash machine" (see Figure 8-2). The digital images, incidentally, are sent back to city hall by fixed-point wireless.

Figure 8-2
A camera cop on every street corner? Photo by Daly Road Graphics.

Orwell's warning to us did not anticipate a culture in which millions of citizens carry cameras of their own, replete with the ability to transmit pictures to the world. We predicted that eventuality when this book was first being written, but it has come to pass even before the book was finished. In August 2002, Samsung marketed an Internet-ready cell phone with an optional, $70 digital camera attachment. Every cybercitizen will soon be carrying a device capable of communicating what they say, where they are, and what they are seeing anytime anywhere. That will change us as profoundly as the arrival of ubiquitous, tiny, inexpensive computers.

One hint of how it might affect us may be the Rodney King incident. In March 1991, a speeding motorist was stopped by several police officers after refusing to pull over. A neighborhood resident, George Halliday, used his HandyCam to videotape the officers from his balcony as they hit King's prostrate body 56 times with their nightsticks, causing kidney damage, brain damage, and several skull fractures. By the end of the next day, the 81-second tape was on TV screens worldwide. Some of the officers were indicted for filing false reports, then acquitted, and then reprosecuted federally after riots in Los Angeles had left 54 dead, 2,300 injured, and 13,000 in jail. Damage to Los Angeles County was estimated at $700 million. Nobody can deny the power of Halliday's pictures. For their part, police departments across the country have taken to installing cameras in their patrol cars.

The chief social effect of the Internet is its implicit democracy. An individual with a web page can disseminate a message as widely as a large corporation, government, or broadcasting network, because on the Web, all present an equal face. Web data zips across oceans and languages, short-circuiting humanity's six degrees of separation. The Internet will inexorably displace newspapers and television because it is interactive. Metcalfe's Law states that a network's worth is directly related to the number of people on the network. Put mathematically, he says that "where *N* is the number of nodes, the power of a network is *N squared*." So thankfully, Orwell was only half right. As millions more users pile onto the global Web, interconnecting with fewer intermediaries, the power of big entities to determine what individuals say and see will become increasingly diffused among billions. The last obstacles to knowledge will fall—the barrier of distance, of national borders, and with the widespread use of Wi-Fi networking, even the barrier of inconvenience.

In the same way that various media have converged onto the digital highway, so we also are on the verge of becoming one people, as the Internet gives us a common medium of exchange. On a grand scale, that emerging global super-culture will control us in some ways, but liberate us in others. As to the details of how it will play out, we can only say this: No technology, including wireless networking, is good or evil by itself. Intelligent machines are willing servants. They do whatever they are told to do, be it wise or foolish. They work with great efficiency, total commitment, and no conscience. It is up to *We The People* to think first before we give the orders.

Figure 8-3 The author, hard at work in his Wi-Fi office. Photo by Daly Road Graphics.

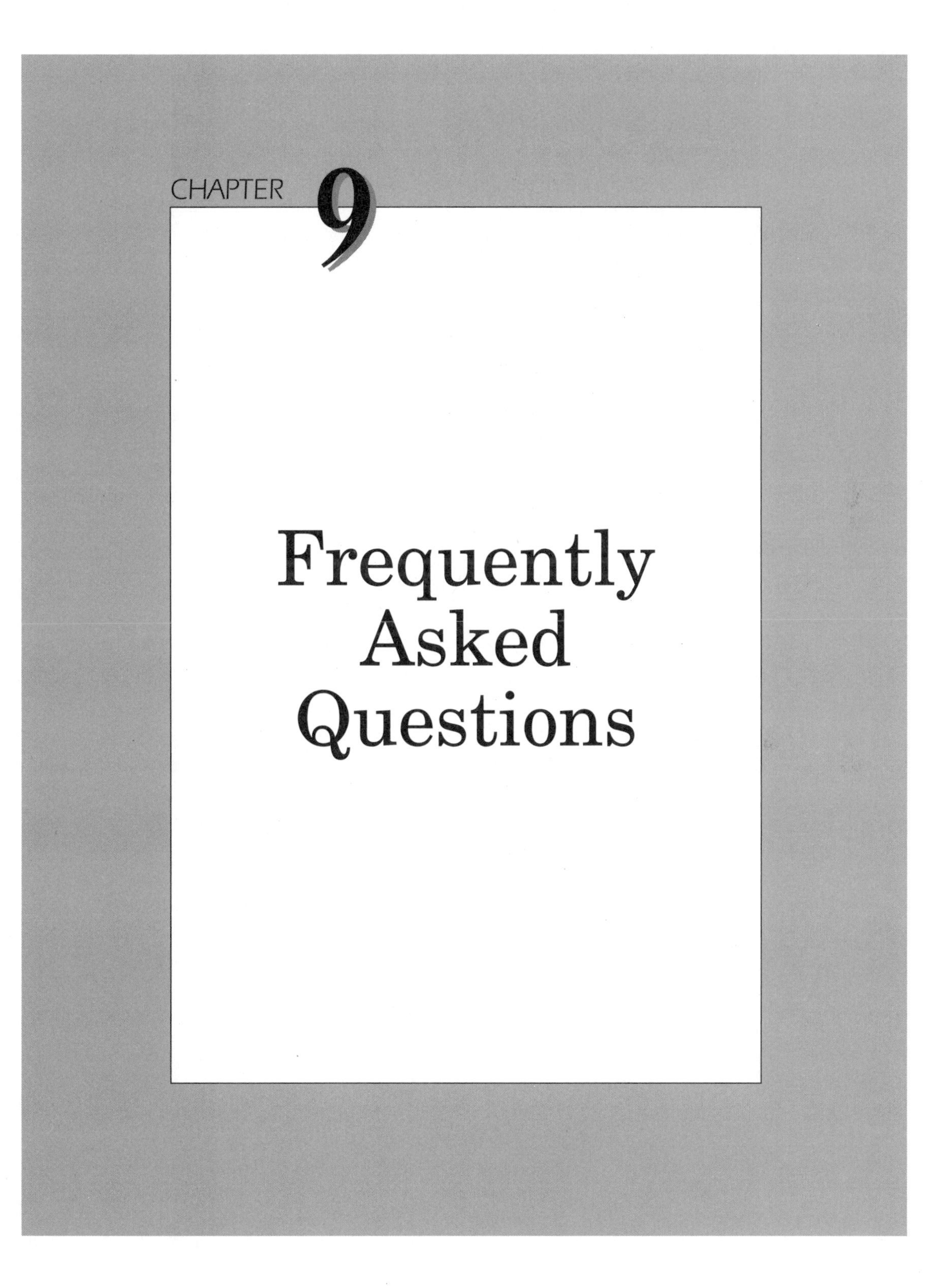

CHAPTER 9

Frequently Asked Questions

In this chapter, we present common questions and shortcut answers. Some of these pages were provided by Wi-Fi manufacturers' tech support teams. Those in the section that follow came from friends and co-workers, email queries, plus some taken from equipment manuals and manufacturers' web sites.

Wi-Fi Networking

Q: What is Wi-Fi anyway?
A: Wi-Fi, or wireless fidelity, is a name for radio-connected computer networking. It is similar to *high fidelity*, the name given to the first stereo phonograph records.

Q: Why is Wi-Fi important to me?
A: It gives you the ability to share files, pictures, sounds, and email between computers without running cables. It enables you to move around and stay connected. It enables you to set up a network with other users just by being in the same building with them. It enables you to access the Internet from a mall, a coffee shop, a conference room, or your living room with equal ease.

Q: How much is this going to cost me?
A: It all depends. If you've purchased a *personal digital assistant* (PDA) or a laptop in the past year, you may already have it built in. If you want to install it in an existing PC or laptop, prices for a single Wi-Fi card range from $40 to $140. If you want to connect your entire home's PCs to the Internet, you'll need a base station for $130 to $500. Of course, you'll have to get a high-speed link from your house to the Internet first, and prices for that can vary. (Refer to Chapter 1, "Is This Trip Necessary?")

Q: How can I justify that kind of expense?
A: The same way you justify the expense of a telephone or a cable TV line. If you work from home or have a home-based business, an Internet-linked network is a necessary business expense. Wireless nets are a great deal cheaper than running new cables, and the infrastructure can be relocated.

Q: If I put a wireless card in my PC, can I use it at work too?
A: New PCs have them built in at the factory. If your corporation has a wireless net, yours will work with it because all Wi-Fi hardware adheres to a common standard. Interchangeability is the beauty of Wi-Fi. Check with your firm's IT manager first, however, to make sure. You will probably need some special passwords and keys to get in. You can also use Wi-Fi laptops to pick up and drop off information at Wi-Fi customer sites.

Q: Does Wi-Fi work with any other 802.11b equipment or brand, such as Lucent, 3Com, or Intel? If my router is from one brand and I have a different brand wireless network card, will they work together? I see 802.11b on the equipment boxes. Why is that important?
A: The 802.11 specifications are a *wireless LAN* (WLAN) standard developed by the *Institute of Electrical and Electronics Engineers* (IEEE). They specify a radio interface between a wireless client and a base station, as well as among wireless clients. If the box label also says Wi-Fi, then it has been tested as being compatible with any other device bearing the Wi-Fi label. It may even be compatible otherwise, but without that label, you can't be sure.

If you mix equipment from different vendors, be certain that you set up your *network interface cards* (NICs) and the router with the same *Service Set Identifier* (SSID), and set them to infrastructure mode. Also, if you have *Wired Equivalent Privacy* (WEP) enabled on the router, you need to enable it on the network cards as well and use the same encryption keys.

Q: I can use my PDA's infrared port to sync data with my laptop. Why can't I use Wi-Fi for that?
A: You can. If your PDA is Wi-Fi compatible (most of the new ones are), then you can use it with laptops or other Wi-Fi-capable PCs. If your PDA isn't so equipped, you may be able to retrofit it with an adapter from the manufacturer or with an add-on peripheral card. Wi-Fi adapters using the *compact flash* (CF) form factor are widely available, for example.

Q: If I can connect my PC to somebody else's, then what's to keep them from connecting to mine even if I don't want them to?
A: Good question! You have to do some configuring to determine in advance who gets in and who doesn't. Mobile Wi-Fi terminals can be set up to seek out and connect to any nearby open systems. You have to set up a unique system identifier and take other steps to assure that your PC isn't open. One is to use WEP, which will encrypt the signal so that only the computers with same encryption key will be admitted to your access point. Another feature you can use on your access point is to allow only the *Media Access Control* (MAC) address of your own wireless adapters to connect and to disallow unknown MAC addresses.

Q: Can I walk out of the building (or home) and still keep working?
A: Yes, but look where you're going! Seriously, the question should be, how far can I walk? The maximum range for these systems is quoted as 300 to 600 yards, if nothing obstructs the antennas. More realistically, Wi-Fi is sometimes called a two-wall technology. You can probably work from your

backyard or perhaps even across the street. It depends on the environment and even the weather. (High humidity tends to weaken the signal.) As the signal degrades, the link slows down.

Q: Can I have more than one access point on my network? What happens when I move from one to the other?
A: Yes, though you probably won't need more than one in a home environment. If you move farther from one and closer to another, at some point you will be seamlessly handed off in the same way a cell phone conversation is passed along from tower to tower when you talk from a moving car.

Q: Is Wi-Fi like those data-capable cell phones I've seen on TV?
A: Not exactly, not yet. Cell phones are being introduced that will use Wi-Fi technology for cheap, high-speed access from public hot spots when they are within range of one. Presently, however, cell phones use their own network, which is slower and more expensive. But they do work from anywhere.

Q: What's a hot spot?
A: It's the Wi-Fi equivalent of a public telephone. Any location where the public is invited to access the Internet through a wireless base station is a hot spot. Examples are airport lounges and coffee shops. So many exist now that directories to them have been set up on the Web. Most charge for the privilege through some kind of prepaid plan, but some freenets cover city parks in major cities. You can sit on a bench there, feed the squirrels, and check your stocks for free.

Q: I'm still not sure. Is this just a passing fad, like hula hoops?
A: Wi-Fi has arrived and is here to stay. Apple is building Wi-Fi into their Macs at the factory, and in mid-2002 Microsoft began to market their own named brand. Businesses without a customer hot spot will soon be regarded as second class, like businesses that don't take credit cards. Homes without fast Internet wireless service will be regarded as rustic, like a cabin in the woods without running water.

Q: If I buy into this stuff, what guarantee do I have that it won't be obsolete in a year?
A: After initial failed attempts to go it alone, manufacturers learned the hard way that they could sell more hardware if it all worked together. In the same way, they know they can sell more new stuff if it works with the old stuff, so they try to make it all backward compatible. Even so, everything except good manners becomes obsolete eventually. Take comfort because you will get your money's worth; it won't get old soon, and the new stuff really is better.

Q: Let's say I set up a Wi-Fi network in my apartment. What's to stop me from deliberately letting my next-door neighbor on my network, so they can get to the Internet?
A: Technically, nothing, if your neighbor is within range. Your *Internet service provider* (ISP) probably has a few thoughts on the matter, because the price they quoted to you didn't include a hoard of mystery passengers. Some ISPs don't mind "piggybacking." Some will even sell you the equipment. Most just mumble and look the other way when they are asked about it. You should ask anyway before you go ahead. If they throw a fit and it's really important for you to be a swell neighbor, then find yourself a new ISP.

Q: What's to keep me from putting an antenna on top of my house and going into business for myself with lots of customers (see Figure 9-1)?
A: Again, ask your ISP. Then think about range limitations, your contractual obligations, permits, fees, customer support, security, and billing. Most freenets get away with it because they are free. But where there are profits, there are taxes. If you are still serious about this, check out www.joltage.com. They may be able to set you up as a one-man hotspot.

Q: I heard about cell phones causing cancer? Will this equipment do that?
A: Wi-Fi runs at lower power levels than cell phones. One watt is the maximum, and it's usually considerably less than that. Nobody has attributed health detriments to Wi-Fi.

Figure 9-1 Wi-Fi home networks can provide a competitive edge to a small businessperson. (Photo by Daly Road Graphics)

Q: Do cell phones interfere with Wi-Fi signals? Or vice versa? How about cordless telephones?
A: Older cordless phones and baby monitors can clash, because they all share the 2.4GHz *Industrial, Scientific, and Medical* (ISM) band. Newer cordless models use a different band, have better shielding, or use newer technology that avoids the same frequency if it is already in use. Older Bluetooth wireless devices can also interfere, as can leaky microwave ovens, police radar guns, experimental ham radios and even electric motors. Cell phones use entirely different frequency bands.

Q: What's the difference between infrastructure mode and ad hoc mode?
A: In infrastructure mode, all the wireless stations communicate through an access point or a wireless router. They don't communicate directly with each other. In ad hoc mode, all the clients act as servers as well in a peer-to-peer network that operates without an intermediary.

Q: What's the difference between an access point and a router? I've heard the terms used interchangeably.
A: An access point interfaces a wireless network to a wired one. Wi-Fi routers do this too, but they do a lot more besides, such as *Network Address Translation* (NAT), switching between wired ports, and sharing a single Internet link to multiple workstations, including wireless ones. You can use a router like an access point, but not vice versa.

Q: What is *Network Address Translation* (NAT)?
A: NAT is the method routers use to prevent hacking into your *local area network* (LAN) and to share one Internet connection with as many as 253 stations on the LAN side of the router. It swaps private, nonroutable *Internet Protocol* (IP) addresses on your local machines for a single public IP address that is visible on the Internet side. Only the router's address is visible to would-be Internet hackers.

Q: Is NAT the same as a firewall?
A: Not exactly. The term firewall is sometimes used to describe the way a router can hide local LAN IP addresses from outsiders. A true firewall is a special-purpose device that checks every incoming packet and filters out dangerous ones based on complex rules defined by a system administrator. That process is called *stateful packet inspection* (SPI). Firewalls can also restrict access to a LAN by requiring logins or *virtual private network* (VPN) authentication certificates. They can also block inside users from seeing particular outside web sites.

Q: When my Wi-Fi *Personal Computer Memory Card International Association* (PCMCIA) card is working, a buzzing noise comes out of my laptop speakers.
A: The radio energy of the card is leaking into the laptop and the speakers because the card cavity wasn't adequately shielded. Unless the laptop manufacturer decides to retrofit it, you're going to have to live with it.

Technical Support FAQs

These questions and answers are the ones most frequently asked of professional Wi-Fi support engineers. These technicians hope that if you see your question here first, you might save yourself a phone call. Those in the section that follows came from Linksys.

Q: If your PC is in ad hoc mode, can you still do roaming to other ad-hoc-mode PCs?
A: Yes. When you roam, your PC's Wi-Fi interface will pick the strongest available signal and then it will connect you to it.

Q: What are some good Wi-Fi channels to use when setting up a *wireless access point* (WAP)?
A: The common channels are 1, 6, and 11 because they won't overlap with other frequency channels.

Q: Why can't I get a signal from my access point?
A: First, set the mode to infrastructure. Then verify that the SSID you've set is the same on both the access point and the remote unit.

Q: I'm getting a strong signal, but I can't pull an IP address from my router/*Dynamic Host Configuration Protocol* (DHCP) server?
A: First, make sure you have the latest drivers. You can check on the support web page and, if necessary, download them from there. If that's no help, click Setup, Advanced, and Wireless, and change the RTS and Fragmentation values to 2304. You also have the option of manually assigning a static IP address.

Q: Does 128-bit WEP encryption work?
A: Mostly. Programs are available that will unravel it for a hacker who is willing to take a day-long sample of traffic. Newer standards are coming that will be harder to crack. WEP should be used along with other provisions if you want your network to be secure.

Q: Can a 802.11b device talk to a 802.11a device?
A: Not directly. These are two different standards running on different frequencies and using different transmission methods. However, some manufacturers are producing dual-band stations with both interfaces included. In that special case, one medium will be cross-translated to the other.

Q: Does VPN work wirelessly?/VPN does not work on my wireless computer, but it does work on my hardwired computer.
A: By itself, Wi-Fi is not a problem for virtual private networking. However, you will have to configure your router/firewall to allow a VPN, and you will need a static IP address to make it work. You may need to upgrade the router with the latest firmware, which is available on the tech support download page. Your company's IT department will have to help you set it up, because VPN can be implemented in several different ways.

Q: Is it possible to change the antennas?
A: That depends on the manufacturer and the model. Many interface cards do have sockets that will accept an extension cord to an add-on antenna. Some of these antennas mount on the ceiling for a better range (doubling is typical) in industrial settings. Some are extremely directional and are used to bridge surprisingly distant WLANs. Most of the time, you won't need to use special antennas.

Q: How many Wi-Fi nodes does a wireless router support?
A: Depends on the manufacturer. Some say 254, but 50 is a more typical maximum number. Remember that the 11 Mbps throughput for a single access point is shared between all the users on that base station. If you have many heavy users, you will have to add access points. They will automatically load balance.

Q: With Windows XP, do you install the software or hardware first?
A: With WIN XP, you should install hardware first. The reason for this is that if you install software first, the drivers *and* the wireless configuration utility get installed. Unless you have the early, first version of the WAP11 access point, you should use the Zero Configuration Utility built into XP. If you install the hardware first, only the drivers will get installed. But that's all that we need.

Q: What can I do to increase the range of my wireless devices?
A: One thing you can try is to change the channel of the access point. A different channel may increase the signal. If you have any 2.4 GHz cordless phones, change the channel of the phone or move the access point to a different location. You should try to place it in the center of your residence.

Also, make sure that other devices or barriers, particularly metallic ones, do not block the antennas.

Q: My Windows XP computer is reporting that the cable of my wireless card is disconnected. Why?
A: Take a look at the XP device manager and verify that

- The Wi-Fi card is properly installed.
- The card was installed with the latest drivers.
- You are using the Zero Configuration Utility included with XP.
- All the interfaces on your net are running with identical encryption keys.

If all fails, remove the drivers of the Wi-Fi interface and reinstall them.

Q: How do I uninstall the drivers of my wireless adapter?
A: Insert the CD-ROM that was included with your wireless adapter. This will bring up a splash screen. Then double-click uninstall. If you've disabled the CD-ROM's autorun feature, you will need to manually access your CD-ROM's contents by clicking the My Computer icon, the CD-ROM icon, and then clicking OPEN to find the uninstall utility. Then double-click that.

Q: I have a router, access points, and network cards. Can all of these and my wired PCs get onto the Internet at the same time?
A: Yes. They will all work at the same time. Remember that the units accessing the Internet will have to share the bandwidth of the Internet link. They may slow down.

Q: Can I bridge my Wi-Fi router with my WAP?
A: No. For bridging, you need a pair of WAPs.

Q: I have a lot of printers in my home. Can I have more than one Wi-Fi print server on the same LAN?
A: Yes. Just make sure that when you configure each printer, you point it to the right print server and also that each printer gets a unique port name.

Belkin Components Wi-Fi Tech Support FAQs

Q: How is it that we are able to connect even when the SSID is not the same on all our PCs?
A: The SSID is the wireless network's name. It is much like the workgroup name in Windows networking. If the SSID of your access point or wireless router is WLAN (our default), then each client will associate with it.

The default for our PCMCIA card is to connect to an SSID named ANY. This means the access point does not look for a specific SSID. If you change that to a specific name, something other than ANY or WLAN, then the card will not connect to WLAN because it is being told to connect to a particular access point with a specific name.

Q: What is the difference between 802.11 ad hoc mode and plain ad hoc?
A: Plain ad hoc is an Intersil (the chipset manufacturer) feature. Wi-Fi ad hoc enables multiple computers to communicate in a peer-to-peer fashion. Plain ad hoc enables only two computers to communicate for security. We advise our users to use 802.11 ad hoc mode for peer to peer.

Q: When I try to connect to my access point, I get an error message that says the connection failed/I lost my password to the access point.
A: The default password for the access point is case sensitive, so be sure that you are typing in the correct password in the correct case. If you are typing the password correctly, then perform a reset of the access point. Here's how:

1. Find the small hole on the bottom of the plastic enclosure. It is labeled *Reset*.
2. Disconnect power from the access point.
3. Using a paperclip or small wire through the hole, depress the Reset switch.
4. Hold the Reset switch down while you reconnect the power.

This will restore all the default factory settings. The default SSID for the AP is WLAN and the password is MiniAP (case sensitive).

Q: Does the Belkin access point require a crossover cable or a straight-through cable?
A: You will need a straight-through patch cord if you are connecting to a standard RJ-45 Ethernet port on a switch or hub. If you are connecting to a Belkin router or a five-port switch, their ports automatically uplink, so either type of CAT5 cable can be used.

Q: Can a WAP be used to connect a PC to a wireless network using its existing Ethernet port or NIC?
A: Yes. You will need a CAT5 crossover cable to connect the WAP to the PC Ethernet port. The limitation to this is that only other wireless computers would be able to communicate with the WAP (and the computer that you connected to the WAP). WAPs cannot talk to other WAPs, so the PC that you connected to the WAP would not be able to talk through it to another WAP or to a wireless router.

Q: Can a WAP be used as a repeater or to extend the range of a wireless router?
A: No. A WAP cannot communicate with another WAP wirelessly, and the same goes for wireless routers.

Q: I already have a wired network and a wired router. Can a WAP be used with the wired router to connect wireless computers to the network?
A: Yes. Afterward, your wireless PCs will use the router just like wired ones.

Q: Can I use WAP to extend the range of wireless computers if they aren't already using a WAP?
A: Yes. If two or more Wi-Fi PCs are talking to each other directly in ad hoc mode, you can put a WAP in between. All the PCs will have to be switched to infrastructure mode in order to use the WAP as an intermediate relay station.

Q: When using the wireless *Universal Serial Bus* (USB) network adapter with the wireless router I am unable to access the network.
A: You have two options. One is to rename the SSID in the Wireless settings from the default (WLAN) to another name. If that doesn't work, try uninstalling the wireless USB network adapter(s), rebooting your PC, and then reinstalling the adapter.

Apple AirPort FAQs (Courtesy Apple Computer)

Q: Can Macintosh computers and PCs coexist on an AirPort WLAN?
A: Absolutely. Because AirPort was designed to work in education and business installations as well as the home, AirPort is compatible with Mac systems and PCs alike. AirPort is designed to meet the IEEE 802.11b standard for WLAN products. So, whether you want to join a PC-based WLAN or host PCs on your Macintosh-based WLAN, AirPort brings everything together.

Q: Why did Apple add a second Ethernet port to the base station?
A: For home users, not only does the second Ethernet port give you the ability to share your single Internet connection with your wired computers as well as your wireless ones, but it also puts them all behind the protection of the built-in firewall. In schools and businesses, the second port is an excellent way to extend an existing network or create a new one.

Q: Which of the new features are available to existing AirPort customers through a software upgrade, and which are available only with the new AirPort base station?

A: Upgrading your system to AirPort 2.0 software will provide additional functionality to both your card and your base station. Your AirPort card will be upgraded to support 128-bit encryption and will be compatible with Cisco access points using the *Lightweight Extensible Authentication Protocol* (LEAP). Customers using the original base station will receive *America Online* (AOL) compatibility. Although AirPort 2.0 can upgrade your existing AirPort card to support 128-bit encryption, it cannot update the original base station, which will remain at 40 bits. To host a 128-bit encrypted network in order to use the *Remote Access Dial-In User Service* (RADIUS) authentication protocol or to have a second Ethernet port, which is protected by the built-in firewall, you will need the new base station.

Q: What is required to set up a wireless network using AirPort?
A: Apple has made it easy for anyone to set up AirPort. You need just four things:

- An AirPort-ready Macintosh (iBook, iMac, PowerBook G4, or Power Mac G4)
- An AirPort card
- An AirPort base station
- AirPort software, which comes with the card and base station

You can easily install the AirPort Card in an AirPort-ready computer yourself and have it up and running in minutes. (1) Simply use the AirPort software to set up the base station and the AirPort network. The AirPort Setup Assistant software walks you through the setup process and automatically configures your computer and base station for Internet access. Note that although Wi-Fi-certified wireless PCs are compatible with the AirPort base station, a Macintosh is required to run the software and complete the initial setup.

Q: Is AirPort compatible with America Oline (AOL)?
A: Yes. AirPort is now compatible with AOL. AOL functionality is currently limited to the United States. (2)

Q: Can I share a printer among computers on an AirPort network?
A: Apple offers the same printer-sharing capabilities for wired and wireless connections. In the office, in the classroom, and on campus, Ethernet printers can be shared on the AirPort network just as they are on your network. Two additional options are also available:

- You can use a wireless print server, such as the HP wp100 available through the Apple Store.

- Because AirPort enables the easy exchange of files between computers, home users can simply transfer files wirelessly to the computer that is connected to the printer.

Q: How secure is an AirPort WLAN?
A: New security features such as 128-bit encryption, RADIUS, and a built-in firewall have been added to AirPort to bring more security to your AirPort wireless network. As with a wired network, however, customers with extremely sensitive data should consider additional security measures, such as a VPN, which may be used to further secure their data. A VPN enables users to access their private networks via a secure portal.

Q: Now that the AirPort base station supports 128-bit encryption, will it still be compatible with my existing AirPort network?
A: Yes. It supports both 40-bit and 128-bit encryption.

Q: What is RADIUS, and why is it important?
A: The Remote Authentication Dial-In User Service (RADIUS) is an access control protocol that authenticates users via an authentication database located on a remote server. Network administrators will appreciate the fact that user access lists can be added to or deleted from one place and that changes can be implemented without power-cycling the network. Schools and businesses that want to protect their network from after-hours access can simply shut down the remote authentication server.

Q: What is Cisco's LEAP protocol, and why is it important?
A: Cisco's Lightweight Extensible Authentication Protocol (LEAP) is a user authentication protocol used by some customers with Cisco access points. Cisco access points are popular with many higher education institutions. Previously, only Cisco 802.11 cards were compatible with this security method. AirPort clients are now compatible with Cisco access points using LEAP.

Q: How do I enable AirPort firewall protection?
A: It's on by default. By configuring your AirPort base station to share a single address, the base station will provide private addresses to all the computers connected to your AirPort network (both wireless and wired). Since these addresses are private and known only to your base station, all Internet traffic must be routed through the base station, protecting your computers from most Internet-launched attacks.

Q: Can I transfer files between AirPort-enabled computers without using an AirPort base station?
A: Yes. You can transfer files or play multiplayer games directly between AirPort-enabled computers. To do this, just create a computer-to-computer

network. In Mac OS 9, use the AirPort Control Strip module on both computers to switch from using the AirPort base station to using computer-to-computer mode. In Mac OSx, use the AirPort system status menu (located on the menu bar) to create a network. Once your wireless network is established, use file sharing to share files as you would on any wired network.

(1) Wireless Internet access requires an AirPort Card, AirPort Base Station, and Internet access (fees may apply). Some ISPs are not currently compatible with AirPort. Range may vary with site conditions.

(2) Compatible with AOL 5.0, U.S. only. Simultaneous sharing of an AOL connection requires multiple AOL accounts.

Apple Wireless Technology FAQs (Contributed by Apple Computer)

Q: How does the AirPort base station share a single Internet connection?
A: The AirPort base station shares a single Internet connection among several computers using a protocol called *Network Address Translation* (NAT). The base station assigns private addresses to each of the computers on your network and routes network traffic to them. Similar to the way the post office delivers mail to your home address, the base station takes the information requested from the Internet and automatically sends it to whichever computer requested it.

Q: How many people can use the AirPort wireless environment at once?
A: An AirPort wireless network can now support up to 50 users simultaneously. To ensure that you get the most from your AirPort experience, Apple suggests limiting the number of users to match the available bandwidth and usage demands.

Q: Can AirPort coverage be extended to provide wireless service to a larger area?
A: Although a single AirPort base station can cover a radius equivalent to half the length of a football field, you may want to extend wireless service across a large corporate campus or even an entire school. By placing multiple AirPort base stations in strategic locations on your network, you can cover a large area quite easily. What's more, AirPort software automatically handles roaming from one base station to another. To the user, it seems as if one giant base station is doing all the work.

Home Broadband FAQs

Here are questions and answers about high-speed Internet service to your residence, as collected from our email and manufacturers' web sites. The following questions deal with broadband over cable.

Q: How fast is it?
A: These things are measured with a rubber yardstick. Speed is comparable to Telco DSL, perhaps faster. But it can be slower when more people start using it, particularly on the upstream path, because they all share the line. Most of the time the provider's speed limit is an optimistic estimate based on how fast the cable modem will go on a test bench. In real life, a chain is only as fast as its slowest link.

Q: So what's better, cable or DSL?
A: That can't be answered in a paragraph, because so many other factors are involved, and only a few of them are technical. What's cheaper in your area? Do you already have cable TV? Does the cable company sell two-way data?

Q: Give me an estimate. How much does cable access cost?
A: Fifty dollars per month, give or take a lot. You may get a discount if you buy your own cable modem. You will pay more if you aren't already a subscriber to the TV offerings.

Q: If I'm surfing on my cable modem, can somebody else watch TV?
A: Sure. Be careful about using your own splitters to drive several TV sets, however. The cable modem signal will suffer first if the signal pie is cut into too many slices, and most splitters won't pass a signal in both directions.

Telephone DSL FAQs

Q: I already have a dial-up. What do I get by paying four times as much for DSL?
A: If all you use it for is picking up email once a day, you won't get much. If you want streaming media such as pictures, sound, and web pages that snap instead of crawling onto the screen, then you'll want DSL. Your connection stays on all the time. There's so much bandwidth that you can share it with every PC in the house. (Wi-Fi makes that very easy.)

Q: I'm trying out my new line, and file downloads start out really fast and then slow down. What gives?
A: It just looks that way because of the manner in which your PC calculates speed. Throughput will vary a little during a long download due to crowding or other factors, but not by much. If the average rate is considerably lower than promised, then you do have a problem.

Q: My download rate just went way up (or way down).
A: If the change seems to be permanent, look first for a local cause, such as interference. Have you added any new telephones, particularly cordless ones? If you can unplug and reboot everything, and the throughput change appears to persist, call your ISP and ask about it. Sometimes they remotely adjust the *Digital Subscriber Line Access Multiplexers* (DSLAM) that your DSL modem attaches to in order to make actual throughput match their records. The question then becomes, does it now match what you originally paid for?

Uninterruptable Power Supply (UPS) FAQ (Courtesy CyberPower Systems)

Q: Why does my unit beep and click when no power outage is taking place?
A: It's doing its job, compensating for an otherwise unnoticed sag or surge in voltage.

Q: How does the surge suppression work?
A: Surges are absorbed by a metal oxide *varistor*. The varistor passes only safe levels of current and any excess is absorbed by the varistor and released as heat. As a varistor absorbs successive surges, it slowly degrades until it loses its surge suppression capabilities. That's why you should replace your surge suppressor power strips every few years.

Q: These come in all sizes. How can I tell how large a UPS to buy?
A: UPSs are rated in *volt amps* (VA). If you must know, the formula is watts $\times$ 1.82 = VA; amps $\times$ 120 = VA. Calculate the total VA of all the equipment that you plan to support with the UPS, and then purchase one with a VA rating that exceeds that number.

Q: Why can't I plug my laser printer into my UPS?
A: Every time a laser printer prints, it produces a large surge, which ends up feeding back into the UPS unit. Laser printers degrade our UPS and lead to a failure of internal components. Over time these surges will cause the UPS to malfunction.

Q: Why does my UPS beep when I turn on my monitor?
A: When a self-degaussing monitor is turned on, it discharges large capacitors in order to create an electromagnetic pulse that clears up distortions in the monitor's display. The discharge of these capacitors creates a short powerful surge that the UPS detects and reacts to. Self-degaussing monitors should be placed on a separate surge strip or else be left turned on and allowed to go into sleep mode.

Q: What is the difference between a surge protector and a UPS? Why can't I just use a surge protector?
A: UPS units are superior to surge suppressors for several reasons. Although both can handle surges, only a UPS can provide battery backup power during a blackout to give you the opportunity to save your data. With a surge protector, during a brownout your computer power supply will still have to work harder to make up the lost voltage. A UPS provides constant voltage at the AC outlet.

Q: What does the PC software do for me?
A: In the event of a power outage, the bundled PowerPanel Plus software automatically logs and saves all open files, and then shuts down the computer system in an orderly manner, even if you are not near your computer.

Q: Will a UPS *really* protect against lightning surges?
A: Yes. CyberPower units are designed to stop even the most powerful surges.

Q: Will an *automatic voltage regulators* (AVR) model help to prevent the "blue screen of death"?
A: Yes. Power sags are the second leading cause of crashes. Poor-quality power causes disk errors, which can lead to crashes. Our units automatically detect sags or spikes and buck or boost the voltage to a safe 120 volts (+/−5 percent).

Q: Do I need to run my cable modem through the CyberPower unit?
A: Absolutely. Surges are not confined to your wall AC outlet. Lightning surges can enter your home through several media, including your coaxial cable line. Our surge protectors are designed to suppress these spikes.

Q: Should I unplug my UPS if there is a lightning storm?
A: There is no need to. Suppressing surges is what CyberPower units are designed to do.

Q: What outlets (battery or battery/surge) should I plug my equipment in to?
A: Your computer and monitor need to be connected to the outlets labeled battery/surge. If you have externally powered storage devices such as a tape or Zip drives, they should also be connected to the outlets supplied by battery power. Other peripherals can be connected to the surge-only outlets.

Q: My UPS beeps when the power goes out. Why?
A: This alerts the user to the outage. If the control software is installed, it will automatically shut down the computer for you. If you choose to continue working during the outage, you can turn the control software off.

Q: We live out in the country where the power fluctuates often and the lights dim. Will a UPS help?
A: Yes. Often, rural homes have unstable power, which can be detrimental to computer components. A low-voltage incident will cause the computer's power supply to work twice as hard to produce the same output. This can reduce the life of internal components. A high-voltage incident can permanently damage integrated circuits.

Q: I've replaced several modems since I purchased my computer. Would a CyberPower UPS protect my modem?
A: Absolutely. Modems are very sensitive components. If you access the Internet using a dial-up service, such a unit will suppress or block surges that enter through the telephone line.

GLOSSARY

.gif An image file using the vector-based Graphics Interchange Format.

.jpg An image file encoded and compressed using the JPEG technique.

.mp3 A sound file encoded and compressed using the MPEG technique.

10BaseT Ethernet running at 10 Mbps, originally on telephone or CAT3 wiring.

100BaseT Ethernet running as fast as 100 Mbps on CAT5 wiring. Backward compatible with 10BaseT.

3DES 3DES is a variation on the *Data Encryption Standard* (DES) that uses a longer 168-bit key. It is considered by security experts to be unbreakable (for now).

3G Third-generation technology. Cell phones use it to become mobile data terminals.

56K The top speed (56 Kbps) for a telephone modem using the v.90 standard. Your results may vary.

abandons Customers on hold for so long that they hang up before the Muzak stops.

AC Alternating current. Electricity that changes polarity at intervals, such as the power in your home, as opposed to *direct current* (DC), like the electricity in a battery or your car.

access control list A list of *Media Access Control* (MAC) addresses predetermined as permitted to use a router.

access point A Wi-Fi-to-wired-network interface. It can also connect wireless clients to each other, if they are set to operate in infrastructure mode, or to a broadband modem.

adapter card Hardware, often a printed circuit card that interfaces a computer with the outside world. An example would be a *network interface card* (NIC).

ad hoc mode A Wi-Fi network architecture in which every node connects on its own to every other node, forming a peer-to-peer net, with no intervening equipment.

ADSL, ADSL Lite Asymmetric Digital Subscriber Line. A broadband Internet interface using copper wire, typically sending to the customer at 1.5 to 9 Mbps, and taking from the customer at a slower rate of 16 to 800 Kbps. The lite version is slower.

AES Advanced Encryption Standard. The new replacement for the *Data Encryption Standard* (DES). AES uses a symmetric or private key algorithm.

AirPort The trade name for Apple's Wi-Fi peripherals.

AirSnort A Linux-based program that cracks encrypted Wi-Fi traffic.

AlohaNet The first wireless network, created at the University of Hawaii in 1970.

analog **1:** An electrical signal or waveform in which the amplitude or frequency can vary continuously, instead of in discrete steps. **2:** Older technology in which information is reproduced via inexact copies.

antenna diversity By sending the same signal from two or more points at once, the probability of dead spots in the signal pattern are considerably less. The probability of multipath fading increases.

API Application programming interface. Predetermined software routines that can be called by a program to access a network or other services.

AppleTalk Apple's proprietary networking protocol. It used CAT3 (telephone) wire at 230 Kbps.

ARP Address Resolution Protocol. TCP/IP Interior Gateway Protocol for dynamically converting Internet addresses to physical hardware addresses on LANs.

ARPA, ARPAnet Advanced Research Projects Agency, founded in 1957. Thanks to Bob Taylor, ARPA became the home of the prehistoric Internet.

ARQ Automatic Retransmission Request. An error correction scheme in which the sender mathematically analyzes the payload data and appends a checksum number. The receiver calculates a checksum based on the received payload. If the two checksums match, the receiver sends an ACK, or acknowledgement. If not, a NAK is sent and the original data is retransmitted until the two checksums match.

asymptotic limit As a society becomes increasingly saturated with new technology, it becomes less capable of effectively absorbing more (cybernetics).

ATM Asynchronous Transfer Mode. The latest high-speed, high-capacity data transmission technology. It moves voice, video, and data simultaneously, encapsulating it in fixed-length (53-byte) packets called cells.

attenuation The loss of signal power, usually over distance. Causes include absorption, deflection, reflection, and diffusion. Measured in decibels, or db, like its opposite, *gain*.

attenuator A device that intentionally weakens an electrical signal.

authentication The process for validating a user by checking his or her login (username and password) against a list of authorized users.

autorun The Windows feature that automatically starts and runs a program on a CD-ROM after it is inserted.

backward compatible A new technology that adds capabilities while still working with earlier versions.

bandwidth A section of radio frequency spectrum, or the amount of data that can pass through a channel. For analog devices it is expressed as hertz, and for digital in bits per second.

BBIAB, BBL Abbreviations used in games, meaning I will be back in a bit, or back later.

beaming Copying data from one PC to another using an infrared light beam.

Betamax Sony's attempt to set a standard for home videotape recorders. VHS won the competition, but Sony stubbornly persisted in making the product until August of 2002.

Betamaxed Having your product superceded by another, even if yours is believed to be better.

bidirectional printing Also known as Bitronic. Modern printers report their status (ink levels, paper out, error conditions) and accept data to be printed. Printer interfaces must therefore be two-way communications devices.

bindings Assigning preprogrammed functions to designated "hot keys" on the keyboard (in game usage).

BIOS Basic input-output system. Instructions are executed at boot time that inform the computer how much and what type of hardware it will use, such as memory, hard drives, and so on. It's stored on a special memory chip, called a *complementary metal oxide semiconductor* (CMOS), kept alive by battery during power-downs.

bit Binary digit. Either 0 or 1.

bit rate Transmission speed, a determining factor in multimedia quality.

Bluetooth A wireless protocol designed primarily to replace connecting cords on personal peripheral devices, such as headphones. This short-range application is known as the *personal area network* (PAN). The name comes from a Danish king who liked to eat blueberries.

bridge In networking, a device that connects two LANs using the same protocol.

bridge taps Unused spur lines connected to a subscriber's phone line. One cause of DSL signal attenuation.

broadband Relatively high-speed, high-capacity network connections.

brownout When AC voltage dips 10 percent less than a minimum level. This is unlike a blackout, during which it disappears entirely.

browser Programs that aid in connecting to web sites and displaying their contents.

BTSOOM (used in chat, email, newsgroups) Beats the s*** out of me.

BTW (used in chat and email) By the way . . .

bug A built-in error, usually software. For games, this may open up a cheat, exploit, or other unintentional advantage.

bunny hopping (Games) A tactic in which a player makes a harder target by moving constantly.

byte A series of consecutive bits (eight is typical) that are dealt with as a unit.

cable modem A device that provides a broadband interface to the Internet using a cable-TV line as a medium.

cache A temporary, relatively high-speed data storage bin.

camping To sit on a dial-up Internet connection, whether in use or not, for a prolonged period. In games, this is sitting in one spot, either to guard something or to ambush an opponent.

cardbus A later, nonbackward-compatible version of the *Personal Computer Memory Card International Association* (PCMCIA) form factor, designed for laptop peripheral adapter cards. This features low power consumption with 33 MHz, 32-bit data transfers.

case sensitive An uppercase or capital letter does not equal the same letter in lowercase. For example, with a case-sensitive logon, "John Doe" will work, but "john doe" will not.

CAT5 Category 5. Multiconductor network cables of various gauges capable of a top bandwidth of 100 Mbps. CAT5e (enhanced) supports signaling rates of up to 200 MHz over distances of up to 100 meters.

channel A pathway between two points used to transmit data.

clan An ad hoc collection of online gamers into a league.

Clarke Belt The string of geosynchronous, high-orbit (22,000 mile) earth satellites used to relay global communications. They are aligned over the equator and circle the globe once every 24 hours, thus appearing stationary in the sky above earth stations. The Clarke Belt is named for the author who first proposed the idea, Arthur C. Clarke.

CLEC Competitive Local Exchange Carrier. An unregulated competitor to the traditional phone companies, including ISPs, CATV providers, cell phone companies, and even power companies.

closed system A Wi-Fi security configuration parameter, forbidding access to users with blank identifiers.

CMOS Complementary metal oxide semiconductor. A low-drain, battery-sustained memory chip used to store hardware configuration information for a PC's bootup sequence.

CMTS Cable modem termination system. The equipment at the cable TV provider's offices that interfaces with the Internet on one end and subscribers' modems on the other.

coasters Unwanted CD-ROMs, like the ones that result from spoiled copy attempts or that come in the mail.

coaxial cable A transmission media consisting of a central wire surrounded equidistantly by insulation, then by metallic shielding to prevent signals from leaking in or out, and then more insulation. Cable-TV cable is an example.

cockpit effect Also known as the Lombard Effect. People adjust their speech according to aural feedback. In noisy or echoing environments, they speak loudly or get confused.

colocated When a device in a site is privately owned by someone else, such as a business' web server kept at an ISP instead of the business.

community string A series of characters that functions like a password for remote access and determines which users are allowed what degree of access to networking equipment.

condenser microphones Inexpensive, high-quality microphones that must be powered by the recording device they are plugged into, unlike dynamic or piezo-electric varieties, which translate the kinetic energy of sound directly into electricity.

conditioning Cleaning up a telephone line so it will adequately carry a high-frequency DSL signal the full distance to your residence or business.

connector conspiracy The design of equipment so that it cannot be used with similar equipment from a different manufacturer. This is sometimes achieved through the use of unique cables or plugs.

convergence Now that most media can be transmitted digitally, the distinctions between different types are blurring (converging), as are distinctions between manufacturers and service providers.

cookies Small text files planted in your browser's cache. Originally designed to enable revisiting without relogging in. Sometimes used to track your surfing and other preferences.

CPE Customer premises equipment. Any equipment provided by the subscriber at their site, such as a Wi-Fi access point/router.

crossover cable A special ethernet patch cord that changes wired ethernet sockets from uplink to downlink or vice versa. It works by internally swapping transmit with receive wires. That is, it crosses them over.

CSI Customer self-install. The most common method of setting up a DSL network at a residence, in which a subscriber installs his or her equipment and microfilters.

CSMA/CA Carrier Sense Multiple Access/Collision Avoidance. The IEEE-specified way that Wi-Fi stations (mostly) avoid trampling on each other. The stations listen before transmitting, because unlike wired nodes, they cannot listen while transmitting. If a sender does not get an acknowledgement from the receiver, it will wait a random interval and try again.

cybarian Cybernetic librarian. A researcher who does most of his or her work on the Internet.

cyberspace A virtual universe of users and computers connected by networks. Originally coined by William Gibson in his sci-fi novel *Neuromancer*.

dancing baloney Animated .gif or other files that serve no useful purpose on a web page.

DAT Digital Audio Tape. A reusable media for copying sound and sometimes computer data.

data mining The use of sophisticated statistical analysis to discover patterns in data, often to improve a vendor's understanding of customer preferences and behavior.

day extender A facility or service that effectively extends the number of working hours in a day. A Wi-Fi hot spot in an airport lounge is an example.

deathmatch The simplest form of *first-person shoot-'em-up* (FPS) gaming. Shoot your opponent first.

default A standard parameter that is automatically entered for you, unless you override it with something of your own.

definitions Mathematical descriptions used to recognize dangerous software, such as viruses.

demark The demarcation point at which the service providers' (DSL or telephone) wires connect to your network. Typically, this is the "phone block" at the rear of your house.

DES Data Encryption Standard. A method of concealing data and reconverting it to clear text, designed by the National Bureau of Standards. It uses a block cipher algorithm.

DHCP Dynamic Host Configuration Protocol. A method to distribute IP addresses to client computers at boot time from a central point. The opposite of static or fixed addressing.

disturber A source of noise on a data line that comes from an inductively connected nearby data line.

DMZ Demilitarized zone. The demilitarized zone is router feature, which permits a designated computer to receive Internet traffic unimpeded by the router's firewall protection.

DNS Domain name server. An electronic directory that converts a text-based URL to a numerical IP address.

DOCSIS Data over Cable Service Interface Specification. A set of standards for transferring data via cable TV and cable modems. The Multimedia Cable Network System manages DOCSIS.

domain name An addressing system and naming convention for sites on the World Wide Web. The domain for NASA is nasa.gov. Their domain name is www.nasa.gov.

dongle A flexible, short, wired adapter connecting a PCMCIA card on one end to a larger socket on the other (see Figure G-1). It dangles and snags.

DOS **1:** Disk operating system. An older text-based operating system for PCs. **2:** Denial of service. A hacker attack on a web site by swamping it with phony information requests.

downlink The path from a server to a client or, for example, from the Internet to a home PC. Literally, the path from a satellite down to an earth station.

downstream From the host to the client, as in download.

Figure G-1
A big dongle. This one is for type A USB ports. Photo by Belkin.

drops The wire that drops down from a telephone pole to a customer site. The terminal point of a DSL line outside a home or business. A short-distance LAN cable that runs from a wall block to a PC on the floor.

DSL A digital subscriber line. Often used to provide a broadband link to the Internet.

DSLAM Digital Subscriber Line Access Multiplexer. A device in the central office, which combines data from several subscribers onto a single line for transport to the backbone.

DSSS Direct-Sequencing Spread-Spectrum. A Wi-Fi-compatible radio frequency-hopping technique that spreads data packets over a range of frequencies for privacy and to avoid fixed-frequency interference.

duel In games, two players only.

dummy DMZ A nonexistent LAN IP address deliberately placed in a router's DMZ to create a dumping ground for unwanted incoming Internet traffic.

economies of scale An economic principle that holds that the production cost per unit decreases as the number of units increases up to a point, at which administrative and other overhead begins to eat into the savings. It's the reason that IBM bought DOS from Microsoft instead of writing it themselves.

Emoticon Icon A combination of chat or email text characters used to indicate an emotion, i.e. :-) for a sideways smiley face.

encryption The mathematical processes of converting clear text into unreadable copy and back again. Used to prevent reading by unauthorized persons.

end point A router feature that enables the initiation of VPN tunnels to some other location, either a VPN client or another end point.

Ethernet The most widely used technology to tie computers into a LAN. Invented by Xerox in Palo Alto, California.

experience In role-playing games, a number representing the number of deeds your character has accomplished.

exploit A method of cracking an encrypted transmission. In games, it is using a bug in the game to gain an unfair advantage.

fallback The process by which a Wi-Fi interface slows down to accommodate a weak or distorted signal. Sometimes called rollback.

FDM Frequency Division Multiplexing. A circuit's total bandwidth is sliced into subchannels, each of which can then be used for separate data streams.

FEC Forward Error Correction. A method to guarantee the reliability of a link by cutting a data stream into groups. Individual bits are progressively repeated in successive groups. The receiver can correct corrupted data on the fly. Used in 802.11a, not 802.11b.

FHSS Frequency Hopping Spread Spectrum. A transmission method in which data are sent over a random sequence of discrete ratio frequencies in order to elude unauthorized listeners and dodge single-frequency interference. Invented by Hedy LaMarr and George Antheil.

firewall A hardware or software device that restricts access to a local network from the Internet.

firmware **1:** Semipermanent programming used with hardware and software, sharing the characteristics of both. **2:** Nonvolatile memory, which can be erased, reused, or updated, as with *Programmable Read-only Memory* (PROM).

fixed-point DSL A wireless method of extending broadband Internet access to a community by using a narrow-beam radio signal as the medium. The endpoints are permanently mounted.

flaming The use of offensive personal attacks to express disagreement, usually in chats or games.

FOC Firm Order Commitment. The date at which outside wiring, such as DSL wiring, is scheduled to your *minimum point of entry* (MPOE).

form factor A component's shape and size, including width, depth, socket type, and so on.

FPS First-person shoot-'em-up. The playing field is usually seen through a character's eyes.

frames per second The number of pictures per second your video card can paint on your screen. Determines the smoothness of visual media, games, and sometimes overall PC performance.

freenets Public access hot spots. Anyone can use them to wirelessly connect to the Internet at no charge.

FTP File Transfer Protocol. A method for copying a file from one computer to another. Developed for the ARPAnet, but it is still used today to upload web pages to servers.

full duplex The simultaneous two-way transmission and reception of data. Sent data does not appear on the sender's screen until echoed by the receiver, thus confirming its successful arrival there.

gateway A point of entry from one network to another. DSL gateways extend the 14,000-foot range limit between a residence and a central switching office.

Get a Life. Admonition originally coined by William Shatner on *Saturday Night Live* to Trekkies. Also applicable to Internet surfing or gaming addicts.

GG Good game. Shorthand compliment to an opponent when the match is ended.

gigahertz, GHz A radio frequency of a billion, or one thousand million, cycles per second. A very, very short-wave radio frequency.

GUI Graphical user interface, pronounced gooey. A visual, picture-based method of interacting with a computer, as opposed to one that is text based or command-line-driven.

half-duplex One-way transmission of data, during which none can be received. Wi-Fi transmitters operate in this mode.

HAN Home area network. A network local to your home or residence.

HF Have fun. Polite salutation to an opponent before a game begins.

HomePNA Home Phoneline Networking Alliance. A trade association of manufacturers who seek to standardize and promote the use of telephone lines as a home networking medium.

HomeRF Home radio frequency. An early standard for residential WLANs, it is no longer popular due to its slow (1.6 Mbps) transfer rate.

hose and close A method used by some tech support people who inflate their call statistics by spouting buzzwords and then referring customers elsewhere, such as "Check the cache for c-span viruses and call back (after my shift ends)."

host A computer that stores and dispenses data to other computers called clients.

hot spot A location equipped with a Wi-Fi access point.

hot swappable Equipment (a USB device is a good example) that can be installed or unplugged without shutting down the computer first.

HTML HyperText Markup Language. Invisible, embedded instructions that prescribe to a browser how a web document should look to the viewer.

HTTP Hypertext Transfer Protocol. A specialized language specifying how browsers and servers connect and swap documents with each other.

hub In networking, a device at which wired nodes connect to each other through a common point. Unlike a switch, bandwidth between them is shared.

hypertext Formatting that enables a user to automatically reference another source by clicking a highlighted word in a document.

IEEE Institute of Electrical and Electronics Engineers. An international professional organization that sets standards for LAN equipment and computers.

IGD Internet gateway devices. The Microsoft name for routers.

ILEC Incumbent Local Exchange Carriers. The traditional, regulated providers of telephone and data services.

IMHO, IMO In my honest opinion, in my humble opinion. Polite precedent to an editorial comment delivered in a text-based conference.

inductive coupling The transfer of electricity from one wire to another without physically touching, via a shared magnetic field. The active principle of electrical transformers.

infrared Electromagnetic radiation (light) whose frequency is above that of microwaves, but below that of the visible spectrum.

infrastructure mode Clients connecting to each other indirectly through intermediary hardware.

interactive media Two-way communication in which user choices immediately affect the output of the sender.

Internet The super-network that interconnects other IP-based networks around the globe.

intranet A TCP/IP-based proprietary network, such as might be used by a corporation.

IP Internet Protocol. The technical specification of how information is to be packaged and transmitted on the Internet.

IP address A unique, 32-bit number that identifies a computer on the Internet.

IPsec Internet Protocol Security, the most widely used method to encrypt commercial data for transport over an untrusted network using VPN.

IPv6 Internet Protocol Version 6. The next generation of global (and interplanetary) networking. It increases the number of available IP addresses and also builds in *Internet Protocol Security* (IPsec).

ISM Industrial, Scientific, and Medical. An unlicensed frequency band (2.4 GHz) used by Wi-Fi.

ISP Internet service provider. The agency that actually connects your PC to the Internet backbone and possibly provides other services as well, such as email.

IVRU Interactive Voice Response Unit. The telephone robot that gives announcements and greetings, and redirects (and logs) your call based on your button pushes.

jabber To continuously stream junk packets onto a network by a defective interface.

Java A hardware-independent programming language devised by Sun Microsystems in 1995 to animate web page graphics and do other tricks. The programs are called applets.

JK (Games, chat) Just kidding.

JPEG Joint Photographic Experts Group. A compression technique for still images in which they are broken down into blocks, and the blocks combined. Their file extension is .jpg or .jpeg.

Kbps Kilobits, or thousand bits, per second.

Kerberos A Unix-based user database used for login authentication. Developed by MIT in the 1980s.

KHz Thousand cycles per second. For sound, within the range of human hearing.

killer app A tool or service so attractive that users will buy software or hardware just so they can use it. The killer application for the Internet was email. Spreadsheets were the killer application for early desktop PCs.

lag In video conferencing or games, an annoying pause between transmission and result.

lame In games, undesirable behavior, from a "Llama."

LAN Local area network. A proprietary collection of computers in a small geographic area such as the same building, room, or campus.

latency The amount of time required to send a ping and then to get an acknowledgement.

LED Light-emitting diode. A solid-state light bulb that uses little power, produces little heat, and lasts a long time. Frequently used as equipment status indicators and more recently in traffic lights and turn signals.

LEAP Lightweight Extensible Authentication Protocol. A wireless user authentication protocol used by Cisco access points and Apple base stations.

lease The temporary use, not permanent ownership, of an IP address.

leetspeak The replacement of text characters to create a semisecret language. An example would be L33tsp34k.

Linux An open-source operating system for PCs. Unlike Windows, it is available on the Internet as freeware without copyright. Users can customize it.

loading coil Used on local phone lines to extend their distance (past 12,000 feet) for voice. They restrict distance for DSL data.

low pass filter A device that selectively permits a signal to go through based on its frequency. In this case, a low frequency, as for a woofer loudspeaker.

LPB, low ping ba*rd** A game player with a perceived unfair advantage by virtue of their faster connection. One with a ping time under 30 ms.

MAC address Media Access Control address. A unique identifier for network equipment consisting of the manufacturer's ID number plus a unique serial number. Expressed as a hexadecimal number.

Martian An IP packet that shows up on the wrong network because the router is malfunctioning or the address is corrupted.

Mbps Megabits, or million bits, per second.

Metcalf's Law The power and value of a network increases logarithmically as more users join.

MHz Megahertz, or millions of cycles per second. A short-wave radio frequency.

microfilter A bandpass device used to separate a DSL signal from lower-frequency voice traffic at a customer residence or business.

MIDI Musical instrument digital interface. A standard for connecting musical instruments to computers and for reproducing music (.mid files) encoded in this manner.

MOD Modification. A change, often an add-on.

modem Modulator-demodulator. A device that converts digital data into analog and back again. Most frequently used to convert audio tones so that computers can use a telephone line to communicate.

MPEG Moving Pictures Experts Group. Actually, a series of standards for compressing data into smaller storage spaces and retrieving it. It works by storing only the differences between changed frames. Their file extension is .mpg or .mpeg.

MSO Multiple Service Operators. Cable TV companies are an example.

MTU **1:** Maintenance test unit, a remotely installed diagnostic device. **2:** Multitenant unit, a prospective location for a DSLAM other than a CO, such as an apartment building. **3:** Maximum transmission unit, the biggest packet size permitted for a given protocol.

multipath High-frequency radio is directional and reflects much as light does. A reflected signal can mix with the slightly delayed original, creating constructive distortion, as the two converge out-of-phase at the receiver.

Murphy's Law The pessimistic assertion that "Whatever can go wrong will go wrong." It has many corollaries and additions, such as "It is easier to get into something than to get out of it," a favorite of authors.

N1 Nice one! A compliment to a game opponent.

narrowband Unlike broadband, a channel that moves data at less than 1.5 Mbps.

NAT Network address translation. The translation of IP addresses within an internal network to a different, single IP address known to an external network. Used to share an Internet connection between several nodes and also protect them from hackers.

nerds 1950s term for unpopular, unattractive losers. In the 1990s, it was redefined to describe unattractive, unpopular millionaires who made it big in Silicon Valley.

NetBeui NetBios Extended User Interface. A bridgeable but nonroutable networking protocol originally developed by IBM.

NetBios Network Basic Input-Output system. A fairly standard software methodology used for application programs to talk to or across a network.

netcast Broadcasting or simulcasting to an Internet audience, as opposed to "terrestrial" listeners.

netizen A citizen of the Internet, or cybercitizen.

Netware A client-server network operating system based on the *Internetwork Packet Exchange* (IPX) protocol. Owned by Novell.

newbies A newcomer to cyberspace. Someone unfamiliar with an Internet application.

NIC Network interface card. A PC's hardware interface to the network.

nick Nickname by which a game combatant is known to his comrades and opponents.

NME In a game, a bad guy.

node A waypoint or endpoint on a network, at which information is processed or reprocessed.

NOS Network operating system. Software that controls and drives a network and provides file sharing. Novell Netware is an example.

octet Eight bits. Usually the same as a byte, but technically a byte can be more than eight.

OFDM Orthogonal Frequency Division Multiplexing. A frequency-hopping scheme used by 802.11a equipment. The data stream and frequency band are cut into slices. The most data travels on the frequency channels with the greatest clarity. Less affected by multipath fading than 802.11b's DSSS method.

OMG, OMFG OmiGawd, and so on. Expressed amazement.

overbooking Intentionally signing up more users than a channel can efficiently service.

own, owning, ownage Beating an opponent by a huge amount.

packet Network information is cut into pieces called packets and dispatched separately, sometimes over multiple routes. It is reassembled at the destination. Like a postal envelope, each of the pieces contains source and destination information as well as payload data.

packet loss Also ploss. Irretrievably missing parts of the data stream.

pairgain A digital (T1) trunk that splits into many analog voice lines. It will not support DSL.

patches Updates to software intended to fix bugs or improve performance. Sometimes referred to euphemistically as enhancements or service packs.

PCI Peripheral Component Interconnect bus. Internal peripheral sockets for adapter cards that can move data in and out of a PC at 132 Mbps.

PCMCIA Personal Computer Memory Card International Association. Four standards governing the design of credit-card-sized laptop peripherals, such as memory, modems, and network adapters. Replaced by the cardbus standard.

PDA Personal digital assistant. A limited-use, palmtop computer.

peer to peer A network organized so that all nodes are equal. Servers work as clients and vice versa.

personal area network (PAN) Interconnected devices in a very limited geographical area, typically a few feet from the user. Wireless headphones are an example.

photophone The first known wireless communication was achieved by Alexander Graham Bell, in 1880 using this device. Bell used a beam of light to carry a voice message between buildings in Washington, D.C. (See Figure G-2)

Figure G-2 First Wireless Network in Franklin Square, Washington, D.C. The dedication plaque reads: From the top floor of this building was sent on June 3, 1880 over a beam of light to 1325 L Street the first wireless telephone message in the history of the world. The apparatus used in sending the message was the Photophone, invented by Alexander Graham Bell, the inventor of the telephone. (Photo by Daly Road Graphics).

PGP Pretty good privacy. A widely used program to encrypt, or scramble, file contents. Invented by Phil Zimmerman.

ping Packet Internet Grope, Packet Internet Gopher. A simple interrogation to verify the existence of a station and measure the time of a round trip to and from it, measured in milliseconds. More complex versions (see tracert) disclose the stops along the way.

ping of death A hacker attack using an enormous ping packet that causes the target to overload and crash.

pocket bongo The act of patting all of one's pockets and/or belt-clipped electrical gadgets to see which of them is beeping, chirping, or vibrating (see Figure 8-1).

POP **1:** Point of presence. The location in a given service area where traffic is consolidated before being sent to an ISP. The place where a carrier puts a switch or router allowing for network access. **2:** Post Office Protocol, the standard method for sending email over the Internet.

ports A number that connects an incoming IP packet with software in a PC; in effect, a door to a program. Packets destined for a web browser, for example, are addressed to port 80.

POST Power-on self-test. A series of diagnostics automatically run at startup or boot time.

POTS Plain old (voice) telephone service. No frills.

PPPoE Point-to-Point Protocol over Ethernet. Software that drives an NIC to emulate a dial-up modem in order to log into an ISP and get an IP address.

PROM Programmable read-only memory. A memory chip whose contents persist through a power down, but which can be rewritten.

propellerhead Someone wearing a beanie cap with a propeller on it, like kids in ancient times. Applied to junior techs who impress their bosses with techno talk or by humiliating vendors, as in: "I had the contract until my PowerPoint prezo was submarined by a propellerhead."

proprietary Privately owned. Not compatible with other vendor's equipment or protocols.

protocol Technical rules governing how network data should be arranged, packaged, and routed.

proxy server A gateway application that relays packets between trusted clients and untrusted hosts. Used to selectively restrict access to some web sites by some users, sometimes to load share and sometimes to cache web pages for multiple users.

QoS Quality of service. The IP protocol was not designed with voice or video in mind, because packets are not guaranteed to arrive at a destination serially or in a timely manner. QoS attempts to remedy this by prioritizing multimedia packets and making others wait instead.

RADIUS Remote Authentication Dial-In User Service. The device that checks your user name and password for validity when you log into an ISP or intranet.

RADSL Rate-Adaptive Digital Subscriber Line. Intelligent DSL modems sense the quality of the intervening copper lines and adjust speed accordingly on the fly.

RARP Reverse Address Resolution Protocol. A method to determine an interface's IP address, given its hardware MAC address.

resolution The precision of a measuring device or the quality of a transmitted visual image, measured in pixels or dots per inch. Hi-res images are smoother and more detailed.

respawn Players who show up again after they've been killed.

RJ-11 The type of socket and plug used to connect telephones.

RJ-45 The type of socket and plug used to connect UTP LAN cabling.

RMA Returned Material Authorization. Permission in advance to return merchandise. If you ship something without an RMA number, it will not be accepted at the service center.

router A device that directs the flow of data between IP networks.

RTFM Read the f***ing manual. An impolite, sometimes inadequate response to a how-to question.

sag A brief (one second or less) voltage brownout.

script kiddies Technically unsophisticated hackers who don't know programming. They copy their tricks from others.

secret shoppers Employees who pose as customers to sample quality of service.

security association The set of mutually agreed upon security protocols automatically negotiated for a VPN tunnel.

shared key A Wi-Fi security option that verifies users before allowing them to participate in a wireless network.

sneakernet The oldest and cheapest method of networking, in which data is copied to removable media and hand-carried from one computer to another, presumably by couriers wearing sneakers.

SNR Signal-to-noise ratio. The maximum bit rate of a modem or Wi-Fi interface is determined by its bandwidth, minus noise. Heavy interference slows them down.

SOHO Small office or home office. One or a few employees who conduct business out of their homes or telecommute from home.

spam **1:** The sending of unsolicited email. **2:** Flooding an area with explosives. **3:** Canned meat, primarily pork shoulder in a gelatin base.

spawn point The location on the field where newly arrived combatants appear.

SPI Stateful packet inspection. Firewalls use it to examine individual packets for source and destination addresses and protocols in order to pass or block them. SPI can even prevent denial-of-service attacks on a web site.

spike A sudden, brief, severe increase in voltage.

splitter A device that either passes or restricts based on the frequency of the input signal. Microfilters are frequency splitters.

spyware A kind of Trojan horse program that collects information about your computing habits and quietly reports it.

SSID Service set identifier. A Wi-Fi network's name. For two Wi-Fi devices to communicate, they must be on the same channel and have identical SSIDs.

standard An agreed principle of technique, as determined by committees assembled by various trade and governmental organizations.

static IP An IP address, which is permanently assigned to a computer or resource.

straight-through cable A normal Ethernet cable usually has the pins on one end directly connected to the same pins on the other end or straight through. *See also* crossover cable.

stumbling Looking for open Wi-Fi systems, as in war driving.

submarined Ambushed from behind (or under) cover.

subnet mask A binary number used as part of an IP addressing technique. It identifies a subnetwork so that an IP address may be shared.

surfing Accessing multiple web sites in a nonlinear search for information or entertainment. Shoulder surfing is the act of looking over a user's shoulder to spy a password or *personal ID number* (PIN).

surge suppressor A device to regulate sudden high voltages and (hopefully) protect sensitive electronic equipment.

switch For data, a device that selectively channels packets between ports. Unintended recipients don't see the traffic destined for other ports or suffer unnecessary bandwidth loss.

SysAdmin System administrator. LAN boss, keeper of the passwords, the one with two pagers.

T1 line Trunk level 1. Before DSL, the most common broadband digital data link, at 1.544 Mbps.

TCP Transmission Control Protocol. The method by which IP packets, which travel and arrive separately, are reassembled in the order they were sent. Since 1973, when it was invented, TCP/IP has become the default network protocol for planet Earth.

Telco The telephone company or other provider of local phone service.

three-finger salute The awkward use of multiple fingers and simultaneous key presses (control-alt-delete) necessary to force Windows-based computers to warm reboot.

TLA Three-letter acronym.

TNF Technically not feasible. One reason a DSL provider may cite in deciding to reject you as a customer. The typical cause is a phone line of insufficient quality.

TP Team play. Opposite of duel.

Tracert Trace route. A DOS-based Windows utility that displays the intermediate relay locations between your PC and a distant host. It also shows the amount of time between them.

transients A quick, brief change in load that impairs networks or power distribution.

transparent GIF A web image whose background assumes the appearance of the web page background, thus blending in.

Trojan horse A supposedly beneficial program that has a secret, unwanted function built in, such as a backdoor or virus.

turbo mode A vendor-specific, faster, perhaps proprietary method of communicating data.

ubiquitous So common that it's taken for granted, like McDonalds.

UDP User Datagram Protocol. A TCP/IP simple protocol that exchanges datagrams without acknowledgments or guaranteed delivery, that is, "fire and forget."

UNII Unlicensed National Information Infrastructure. The 5 GHz frequency band, used for the newer edition (802.11a) of wireless networking equipment.

Unix A computer operating system, used mainly on mainframes. Unlike Windows, which is proprietary, Unix is open source and user customizable.

uplink The (upstream) path from a client to a server, such as from a home PC to the Internet. Also, the path literally upward from an earth station to a satellite.

UPnP Universal Plug and Play. A built-in, mutual process for Wi-Fi and other routers, working together with enabled operating systems, to automatically discover and configure each other. This (hopefully) simplifies the router management process by decreasing the amount of user intervention required.

UPS In electronics, a power supply that draws from constantly recharged batteries. It supplies a constant output voltage regardless of the input, until the sustaining batteries are exhausted.

URL Universal Resource Locator. The name of an Internet resource, specifying its protocol, the server domain name, and the location of the item on the server.

USB Universal Serial Bus. Sockets on PCs that enable the easy connection of as many as 63 devices of many types on an instant recognition and plug-and-play basis. The top speed is 12 Mbps.

UTC Coordinated Universal Time. Twenty-four-hour time of day in Greenwich, England, which is also zero degrees longitude. Formerly Greenwich Mean Time, and presently in the military, Zulu Time.

UTP Universal Twisted Pair. Network cable that resists signal leakage by twisting its wires around each other in a precise manner. Sometimes referred to as CAT3 or CAT5, which is the most common medium for wired networks in use today.

UWB Ultrawideband. A high-speed WLAN operating in the 3.1 GHz and 10.6 GHz bands. A prospective channel for wireless feeds to video displays. These frequencies can also be used to find buried objects.

varistor A voltage-controlled resistor used to limit voltage to safe levels.

VoIP Voice over IP. The nascent practice of using the Internet for cheap, long-distance phone calls.

VPN Virtual private network. A means of securing a data stream from interception by encrypting and then encapsulating it inside packets that must cross public networks on their way to private ones. Most VPNs use the IPsec protocol.

Vulcan neck pinch The awkward use of multiple fingers and simultaneous key presses (control-command-return-power on) needed to force Mac computers to warm reboot.

WAN Wide area network. A geographically large network that may interconnect several LANs.

war chalking Notification to passersby of the presence of a freenet, done by a scrawled icon on a lamppost or sidewalk.

war dialing Automated phone calls, originally to discover phones attached to computers, but more recently to discover phones attached to potential customers. This is done to contact a wide range of target addresses and determine which are available for further exploits. Similar to ping sweep.

war driving Cruising reconnaissance for publicly accessible hotspots, left open intentionally or otherwise.

wavelength A function of frequency. The higher the frequency, the shorter the wavelength.

WDSL Wireless Digital Subscriber Line. A method of extending broadband Internet access to a residence or community by using a narrow-beam radio signal as the transmission medium.

web bug Also, Clear Gif, Tracker Gif, 1-by-1 GIF, invisible GIF. A graphic on a web page designed to report the viewer's IP address, the originating web page, and the time of viewing. Often invisible because they are typically only one-by-one pixel in size, and transparent.

webcam An automated still camera that takes pictures at intervals for the purpose of display on a web page.

WEP Wired equivalent privacy. An encryption scheme that (hopefully) prevents unintended recipients from viewing a Wi-Fi data stream. It comes in two levels 64-bit (sometimes called 40-bit because of the key length) and 128-bit.

WEPcrack A Linux-based program that cracks Wi-Fi traffic encrypted using the WEP algorithm.

Wi-Fi Wireless fidelity. A local area computer network using high-frequency radio waves as the medium of exchange. Devices with this label must conform to the IEEE 802.11b standard, as certified by the Wi-Fi Alliance.

Wi-Fi Alliance A trade organization that tests Wi-Fi equipment to guarantee its adherence to standards, and therefore compatibility between different manufacturers. Formerly known as the Wireless Ethernet Compatibility Alliance, or WECA.

Windows The most widely used PC operating system in the world. It comes in many editions, such as 95, 98, NT, SE, ME, 2000, and most recently XP. All have networking capabilities.

WinIPcfg Windows IP configuration reporter utility. Describes the NIC as well as its MAC and IP addresses. It can be used for a forced reload of the IP address.

WISP Wireless Internet Service Provider. A company that sets up and bills for hot spots, which enable mobile Wi-Fi users to connect to the Internet as they roam.

WLAN Wireless Local Area Network. Easier to set up and change than the wired variety, but often slower. Wi-Fi is one such.

WNG Wireless Next Generation. A proposed IEEE WLAN specification that seeks to combine and streamline all the existing WLAN standards.

World Wide Web Another name for the Internet. Specifically, a protocol for information interchange across diverse computer systems, developed in 1990 by CERN.

X-windows A cross-network, Unix-based windowing system.

yoyo You're on your own. The least expensive method of technical support for users.

Zen mail An email message that arrives with no content or completely blank.

Zip To compress a file, thus reducing its size and transmission time. From the program PKzip by Phil Katz.

INDEX

Note: *Boldface numbers indicate illustrations; italic t indicates a table.*

A

B

C

D

E

F

G

H

I

J–K

L

M

N

O

P

Q

R

S

T

U

V

X–Z

ABOUT THE AUTHOR

Raymond Smith has a B.S. degree in Telecommunications and has worked as a News Director and Editorial Writer for a number of radio stations. A resident of Ojai, CA, Raymond was a programmer and field service engineer for several Silicon Valley computer and data network manufacturers. Most recently, he worked as a network administrator for Verizon.

SOFTWARE AND INFORMATION LICENSE

The software and information on this diskette (collectively referred to as the "Product") are the property of The McGraw-Hill Companies, Inc. ("McGraw-Hill") and are protected by both United States copyright law and international copyright treaty provision. You must treat this Product just like a book, except that you may copy it into a computer to be used and you may make archival copies of the Products for the sole purpose of backing up our software and protecting your investment from loss.

By saying "just like a book," McGraw-Hill means, for example, that the Product may be used by any number of people and may be freely moved from one computer location to another, so long as there is no possibility of the Product (or any part of the Product) being used at one location or on one computer while it is being used at another. Just as a book cannot be read by two different people in two different places at the same time, neither can the Product be used by two different people in two different places at the same time (unless, of course, McGraw-Hill's rights are being violated).

McGraw-Hill reserves the right to alter or modify the contents of the Product at any time.

This agreement is effective until terminated. The Agreement will terminate automatically without notice if you fail to comply with any provisions of this Agreement. In the event of termination by reason of your breach, you will destroy or erase all copies of the Product installed on any computer system or made for backup purposes and shall expunge the Product from your data storage facilities.

LIMITED WARRANTY

McGraw-Hill warrants the physical diskette(s) enclosed herein to be free of defects in materials and workmanship for a period of sixty days from the purchase date. If McGraw-Hill receives written notification within the warranty period of defects in materials or workmanship, and such notification is determined by McGraw-Hill to be correct, McGraw-Hill will replace the defective diskette(s). Send request to:

Customer Service
McGraw-Hill
Gahanna Industrial Park
860 Taylor Station Road
Blacklick, OH 43004-9615

The entire and exclusive liability and remedy for breach of this Limited Warranty shall be limited to replacement of defective diskette(s) and shall not include or extend any claim for or right to cover any other damages, including but not limited to, loss of profit, data, or use of the software, or special, incidental, or consequential damages or other similar claims, even if McGraw-Hill has been specifically advised as to the possibility of such damages. In no event will McGraw-Hill's liability for any damages to you or any other person ever exceed the lower of suggested list price or actual price paid for the license to use the Product, regardless of any form of the claim.

THE McGRAW-HILL COMPANIES, INC. SPECIFICALLY DISCLAIMS ALL OTHER WARRANTIES, EXPRESS OR IMPLIED, INCLUDING BUT NOT LIMITED TO, ANY IMPLIED WARRANTY OF MERCHANTABILITY OR FITNESS FOR A PARTICULAR PURPOSE. Specifically, McGraw-Hill makes no representation or warranty that the Product is fit for any particular purpose and any implied warranty of merchantability is limited to the sixty day duration of the Limited Warranty covering the physical diskette(s) only (and not the software or information) and is otherwise expressly and specifically disclaimed.

This Limited Warranty gives you specific legal rights; you may have others which may vary from state to state. Some states do not allow the exclusion of incidental or consequential damages, or the limitation on how long an implied warranty lasts, so some of the above may not apply to you.

This Agreement constitutes the entire agreement between the parties relating to use of the Product. The terms of any purchase order shall have no effect on the terms of this Agreement. Failure of McGraw-Hill to insist at any time on strict compliance with this Agreement shall not constitute a waiver of any rights under this Agreement. This Agreement shall be construed and governed in accordance with the laws of New York. If any provision of this Agreement is held to be contrary to law, that provision will be enforced to the maximum extent permissible and the remaining provisions will remain in force and effect.